CASUALTIE

Personal Histories from the

Chemically Sensitive

CASUALTIES OF PROGRESS

Personal Histories from the

Chemically Sensitive

Alison Johnson, Editor

With a foreword by
Gunnar Heuser, M.D., Ph.D., F.A.C.P.

MCS Information Exchange
Brunswick, ME 04011

MCS Information Exchange
2 Oakland Street
Brunswick, ME 04011

ISBN 0-9675619-0-6

Printed in the United States of America

Contents

About the Editor

Alison Johnson, B.A., M.A., is a graduate of Carleton College and studied mathematics at the Sorbonne on a National Science Foundation Fellowship. She received a master's degree in mathematics from the University of Wisconsin, where she studied on a Woodrow Wilson Fellowship, and is currently a freelance editor for university presses. She has produced three videos on multiple chemical sensitivity. Johnson and her daughters have multiple chemical sensitivity, which developed at intervals years apart in various family members. Through avoidance of chemical exposures, they are now able to function well in normal life circumstances.

Nothing in this book should in anyway be construed as medical advice. It is important to consult your physician concerning any health problem.

Foreword

In this volume Alison Johnson has collected case histories of patients with a history of chemical injury and resultant sensitivity to chemicals in even very low concentrations.

Having seen and evaluated several thousand patients who are now partially or totally disabled from chemical injury, I believe that the public, government agencies, and health professionals should be far better informed of their plight than they have been to date. This information is now available in the present volume, elegantly and clearly presented and edited by Alison Johnson.

Both this volume and the three videos on chemical sensitivity that Johnson has produced constitute a call for help and understanding from family, friends, employers, physicians, and the government. The first video, *Multiple Chemical Sensitivity: How Chemical Exposures May Be Affecting Your Health*, illustrates through moving interviews with over a dozen patients the way in which exposure to toxic chemicals has produced chronic illness and disability. This video also includes commentary from six leading clinicians and academic researchers in the field of chemical sensitivity, and Volumes II and III contain further interviews with these experts.

Finally, patients need to know that scientific protocols are now available to document both chemical injury and chemical sensitivity. These protocols have been and are being presented by me to professional meetings in Australia, Sweden, Germany, and Chile. In Volume III of the Johnson videos, I present portions of this protocol. Copies of the protocol are available upon request from my office.

I congratulate Johnson upon her videos and her successful effort to present in *Casualties of Progress* a very complex subject in a concise and compassionate format.

Gunnar Heuser, M.D., Ph.D., F.A.C.P.

NeuroMed and NeuroTox Associates
28240 W. Agoura Road, Suite 203
Agoura Hills, CA 91301

Abbreviations and Special Terms

MCS—Multiple chemical sensitivity

EI—Environmental illness or a person with environmental illness

CFS—Chronic fatigue syndrome

CFIDS—Chronic fatigue immune dysfunction syndrome

VOC—Volatile organic compound

OSHA—Occupational Safety and Health Administration

SPECT—Single photon emission computer tomography (a brain scan)

Foil-lined surfaces—People with MCS are particularly sensitive to chemicals such as formaldehyde that are given off by plywood or particleboard. Thus they sometimes cover floors, walls, or cabinet shelves with aluminum foil to seal in the formaldehyde fumes that are released from these products for years.

Offgassing or outgassing—The slow release of chemicals from various substances like paint, soft plastics, new carpet, or particleboard. To get an idea of the many sources in an average home, one need only smell the fumes given off by the plastic-coated wires in the back of a TV after it has been on for a half hour.

Acknowledgments

I wish to thank members of my family for their support in this and previous projects. My husband Wells and daughter Clare have read these stories as they evolved through many drafts. Other readers whose comments have been very useful include Jim Azzara, Marilyn Emery, Betsey Grobe, Claire Hersom, Cordie Johnson, Lynn Lawson, Barbara Riss, Joan Rives, Amy Svoboda, Dino Valaoritis, and Diane Weggel. I also wish to express my thanks to Janet Dauble, Ron Dorn, Lynn Lawson, Jean Lemieux, Ann McCampbell, Karen McDonell, Diane Weggel, and Barb Wilkie for contacts and information and to Marcia Cooper for the cover design.

Preface

This book is a natural outgrowth of two earlier projects concerning multiple chemical sensitivity that drew upon my 20 years of experience as a patient myself and as the mother of daughters with MCS. The first project was a survey in 1996 of 351 MCS patients and their experience with 160 therapies. The people I surveyed sometimes sent me a brief account of their illness or spoke with me by phone about what had happened to them. Often in two or three unforgettable sentences, they documented a life that had become a nightmare because of chemical sensitivity.

As I heard more and more stories, it became all too apparent that those struggling with MCS needed a video that would be helpful in educating friends, family, and others of the reality of MCS as a biologically based illness. Hence in 1998, I produced a video titled *Multiple Chemical Sensitivity: How Chemical Exposures May Be Affecting Your Health*. In 1999, I produced two follow-up videos (see Appendix 1).

Many people who have seen the first video have been curious to know more about the people it featured. The present collection of stories has enabled me to make available interesting and relevant material from fifteen hours of raw footage that couldn't be included in the 90-minute video. In addition, this book has given me the opportunity to publish dozens of other stories, each unique, yet all too familiar to those with MCS.

The stories in this collection were obtained in a variety of ways. Sixteen of them are primarily derived from the video interviews, and a few of those people sent additional written material for their stories. The other stories came from written accounts, tape recordings, or telephone interviews. I asked contributors to write a couple of pages or a dozen pages, whatever they thought necessary to describe their experience living with multiple chemical sensitivity. The essential guideline was to use a conversational tone, and to that end I have not edited these accounts into formal language or grammar. I have included one long story from a young woman named Jennifer that largely derives from letters we had received a few years ago from both Jennifer and her mother that documented in detail, as it was happening, the exhausting, day-to-day struggle that so many people with MCS go through in an attempt to find a place to live that they can tolerate. Abner's and Nancy's personal histories and untimely deaths also illustrate the urgent need for funding for organized housing projects for those with MCS. (See the Afterword for further discussion of this important issue.)

To protect the privacy of those who contributed their personal histories, I have in general not included details that would identify them. One exception is Linda's story because the essential elements have already been made part of the public record in a proceeding before the Vermont Human Rights Commission. I have deleted almost all brand names for liability reasons. In any event, alerting readers to the dangers posed by particular brand-name products does not offer them sufficient protection. Although one particular pesticide was mentioned in story after story that I received, if that pesticide is pulled from the market, it will only be replaced by others of dubious safety.

In general, I have not included accounts of various therapies people have tried because individual reports can give a skewed impression of the efficacy or safety of treatments. Until research funds are available to determine the underlying mechanisms of MCS, it will be difficult to decide which therapies are helpful, which are potentially dangerous, and which may deplete the patient's finances with no benefit. The one therapy almost everyone in the MCS field concurs is helpful is avoidance of chemical exposures. (Further information about my therapy survey and the related booklets I published can be found in Appendix 2.)

Multiple chemical sensitivity is a condition fraught with ambiguity. As you read these stories, it will be apparent that in many cases it is hard to pin the development of MCS in a particular individual to one primary cause. Appendix 7 does, however, provide lists of contributors who believe a given kind of exposure (for example, to pesticide, paint, or new carpet) is implicated in their illness. Whatever their occupation, level of education, gender, or age, most of these people are living quite desperate lives. Their chemical sensitivity makes it difficult for them to hold onto a job or to find housing they can tolerate. The strains of this illness can destroy relationships with friends and family all too quickly. Only when MCS has become broadly recognized as a debilitating and ever-increasing health problem, will sufficient efforts be made to provide accommodations on the job or in housing to help make life bearable for those with severe MCS. Only when research money becomes available to unravel this mysterious syndrome will patients have some hope of recovering sufficiently to live a relatively normal life.

Introduction

New chemicals. They're everywhere in our lives now, from air fresheners to fabric softeners to synthetic carpets. Tens of thousands of new chemicals have entered the marketplace since World War II, and most of them have not been tested for toxicity by any government agency. According to a Government Accounting Office document from 1994, "Over 70,000 chemicals are in use in the United States. Although these chemicals are an important part of our economy, they are often toxic and can have adverse effects on human health."

Among these adverse effects is the proliferation of a new syndrome—multiple chemical sensitivity. The term is suggestive of the problem, but even principal clinicians and researchers in the field have never been able to agree upon a precise definition. In an article titled "Multiple Chemical Sensitivity: A 1999 Consensus" that was recently published in *Archives of Environmental Health,* one group of 34 researchers and clinicians proposed the following criteria for the clinical diagnosis of MCS:

1. The symptoms are reproducible with repeated exposure.
2. The condition is chronic.
3. Low levels of exposure result in manifestations of the syndrome.
4. The symptoms improve or resolve when the incitants are removed.
5. Responses occur to multiple chemically unrelated substances.
6. Symptoms involve multiple organ systems.[1]

One of the most distinctive features of MCS is that people who develop the condition begin to react to low-level chemical exposures that never bothered them previously. In most cases, as the illness progresses, the patient reports that more and more substances cause symptoms. People with MCS can have a wide variety of symptoms as the result of chemical exposures, with different patients having different symptoms. A given patient, however, will usually have the same symptom in response to a given exposure, perhaps getting a headache after exposure to paint or getting arthritic pains after exposure to natural gas.

---

[1] *Archives of Environmental Health* 54, no. 3 (1999):147-49. The complete text is available on the following website:
<<http://www.heldref.org/html/Consensus.html>>

Even though researchers do not yet agree on a precise definition for the condition, the stories in this collection illustrate how chemical sensitivity can destroy a productive life all too quickly. Many people with MCS are so sensitive to fragrances that they virtually become prisoners in their own home, unable to go to church, work, classes, or social gatherings because they will react to the perfume, aftershave, shampoos, detergents, or fabric softeners used by others. And to make matters worse, some of those who insist that MCS is a psychologically based illness state that these people are suffering from agoraphobia, or fear of crowds. That's tantamount to saying to a paraplegic in a wheelchair, "Too bad you don't like to walk."

Readers of these stories will soon see that the line between MCS and other chronic illness syndromes such as Gulf War syndrome, chronic fatigue syndrome, fibromyalgia, and sick building syndrome becomes blurred in many cases. Large numbers of patients with these other syndromes report that since becoming ill, they have become highly sensitive to some common chemical exposures like perfume or diesel exhaust that did not bother them in the past. In 1995, Maj. Gen. Ronald R. Blanck of the Walter Reed Army Medical Center, surgeon general of the U.S. Army, stated that "researchers should take a closer look at multiple chemical sensitivity as a possible cause of some of the symptoms [of Gulf War syndrome]" (see Appendix 3).[2]

Prevalence studies for MCS have been conducted in three states—California, North Carolina, and New Mexico. The California study, which was the largest, was based on a phone survey of 4,046 people. According to the report of this study, which was published in the *American Journal of Epidemiology*: "Of all respondents, 253 (6.3%) reported doctor-diagnosed 'environmental illness' or 'multiple chemical sensitivity' (MCS) and 643 (15.9%) reported being 'allergic or unusually sensitive to everyday chemicals.'"[3] These statistics indicate a significant public health problem that deserves urgent attention because severe MCS is such a debilitating or limiting condition.

Why is there so little recognition of MCS and why has so little money been spent to research the syndrome? This issue is discussed by Nicholas Ashford, Ph.D., J.D., and Claudia Miller, M.D., M.S., two of the leading authorities in the field, in their book, *Chemical Exposures: Low Levels and High Stakes*. The *Journal of the American Medical Association* described Ashford and Miller's book as "a stimulating review of the controversy surrounding multiple chemical sensitivities. . . . Clinicians and policy-

---

[2] *Maine Sunday Telegram*, July 9, 1995, A7.

[3] *American Journal of Epidemiology* 150, no. 1 (1999):1.

makers would do well to read and heed the advice of this book."[4] Ashford and Miller decry the lack of research funds for MCS:

> Scientific investigation related to chemical sensitivity is being stymied by scientists and physicians with financial conflicts of interest (e.g., those working for the chemical industry and those acting as defense expert witnesses in legal cases on MCS) who serve on government panels, editorial review boards, and grant review committees. These conflicts generally remain undisclosed.[5]

One of the most respected researchers in the field of chemical sensitivity is William Meggs, M.D., Ph.D., a professor at East Carolina University Medical Center who has published many articles in peer-reviewed journals detailing, among other topics, his research using biopsies to investigate damage to the nasal lining of chemically sensitive patients. When I recently interviewed Dr. Meggs, he stated: "I've spent a lot of time applying for research grants to try to study these illnesses [like MCS] and the role of chemicals in these illnesses, and my grant applications come back with scathing comments [like] 'Don't spend any money on this research because everybody knows this is all psychological.'"

It's hardly surprising that industry doesn't want anyone to believe that chemical exposures could produce a debilitating condition like MCS. The consequences for corporations would be enormous if members of the public increasingly began to wonder if installing new carpet, using pesticides in their house or yard, or buying particleboard cabinets or furniture might affect their health. And imagine the potential liability problems if people could prove that exposures in factories, hospitals, schools, or offices had destroyed their health.

---

[4] James E. Cone, review of *Chemical Exposures: Low Levels and High Stakes* by Nicholas A. Ashford and Claudia S. Miller, *Journal of the American Medical Association* 266, no. 13 (October 2, 1991): 1858.

[5] Nicholas A. Ashford and Claudia S. Miller, *Chemical Exposures: Low Levels and High Stakes*, 2nd ed. (New York: John Wiley, 1998), p. 271. Nicholas A. Ashford, Ph.D., J.D., is Professor of Technology and Policy at the Massachusetts Institute of Technology. Claudia S. Miller, M.D., M.S., is Associate Professor in Environmental and Occupational Medicine in the Department of Family Practice at the University of Texas Health Science Center at San Antonio and is a member of the Department of Veterans Affairs' Persian Gulf Expert Scientific Committee. For a brief summary of the TILT (Toxicant-Induced Loss of Tolerance) theory proposed by Dr. Miller, as well as a short explanation from her of the concept of masking, see Appendix 3.

To understand the power that industry wields regarding MCS, one need only remember that the tobacco industry managed for decades to keep the public from understanding the hazards of smoking by discouraging any research that would show the dangers of smoking and by funding research that would indicate it was safe. If the tobacco industry, which represents a very small fraction of American business, could exercise so much power, it is staggering to consider the influence against validating MCS that is wielded by corporations when almost every business in the United States is significantly involved in chemical use in one way or another. What advertiser would want to run ads on a TV show that raised the possibility that chemical exposures could be creating serious illness?

Another factor affecting the acceptance of MCS is our society's long tradition of viewing much of unexplained illness as being psychological in origin (psychogenic). Tuberculosis was once thought to be associated with a certain personality type. A few decades ago many physicians claimed that children developed asthma because of domineering mothers, and multiple sclerosis was originally considered to be psychogenic.

Ashford and Miller discuss psychological issues related to MCS at length and make the important point that while some MCS patients have psychiatric symptoms from time to time, that does not mean that the illness is psychogenic. An illustrative example is Mad Hatter's disease from the nineteenth century; the Mad Hatters were indeed crazy, but their insanity was caused by the mercury in the felt with which they worked. One study has shown that panic disorder can be precipitated by exposure to solvents in the workplace.[6]

One of the most telling arguments against the theory that MCS is psychogenic and simply amounts to a fear of chemicals comes from animal research. Several recent experiments replicate features of MCS in a rodent animal model so that they too react to extremely low levels of chemicals with debilitating symptoms.[7] These rodents are obviously not influenced by

---

[6] Stephen R. Dager et al., "Panic Disorder Precipitated by Exposure to Organic Solvents in the Work Place," *American Journal of Psychiatry* 144, no. 8 (August 1987): 1056-58.

[7] D. Overstreet, C. Miller, D. Janowsky, and R. Russell, "Potential Animal Model of Multiple Chemical Sensitivity with Cholinergic Supersensitivity," *Toxicology* 111 (1996): 119-34; W. Rogers, C. Miller, and L. Bunegin, "A Rat Model of Neurobehavioral Sensitization to Toluene, *Toxicology and Industrial Health* 15 (1999): 356-69; B. Sorg, "Proposed Animal Model for Multiple Chemical Sensitivity in Studies with Formalin," *Toxicology* 111 (1996): 135-45.

media accounts of the dangers of chemical exposure. This is clearly an area of research that cries out for funding.

Another unfortunate aspect of the psychological approach to the issue of chemical sensitivity is that critics of MCS frequently suggest that "secondary gain" is a strong component of the condition. According to secondary gain theorists, those with MCS are engaging in certain behavior patterns in order to get special attention or because they want others to take care of them. One does not have to read many of the stories in this collection before it is apparent that this suggestion at best is made in ignorance and at worst represents an exceedingly cruel attitude toward people whose illness has in all too many cases cost them their job, their home, their friends, or their spouse. One contributor to this collection who has received remarkable support from family, friends, and employer, for one brief moment dropped his attempt to be cheerful while I was interviewing him by phone and said, "My life is hell." MCS is an illness of devastating loss, not secondary gain.

In Kelly's story, she describes the experience of her brother, who came back from the Gulf War very ill. After his death, she read this statement in his medical records: "Patient appears to be malingering his illness for secondary gain." Kelly reports that she was amazed when she later learned that veteran after veteran suffering from Gulf War syndrome had seen that same phrase used in their records, as if it were being passed down from higher authorities who wanted to deny the existence of Gulf War syndrome.

But it's hardly surprising to see the concept of secondary gain being applied to MCS patients when one considers the following statement made with regard to "Functional Dyspepsia and Other Nonspecific Gastrointestinal Complaints" in the 1982 *Merck Manual*, a reference guide for physicians:

> **6. Removal of the symptom is not always the goal of treatment.** The illness may have such adaptive value to the patient that the loss from giving up the "benefits" may be greater than the gain from relief of symptoms. Pain or suffering may substitute for more distressing feelings of guilt and sadness. The attention and privileges derived from being chronically ill may also be significant. When the patient overtly or covertly resists management, the illness can be assumed to fulfill certain needs.[8]

If this passage indicates the training that today's physicians received in medical school, then will they think that the chemically sensitive patient

---

[8] *Merck Manual*, Robert Berkow, M.D., ed., 14th ed. (Rahway, N.J.: Merck, Sharp & Dohme Research Laboratories, 1982), p. 713.

who reacts to virtually any medication offered is "overtly . . . resist[ing] management"? Only when the medical profession ceases to view MCS as a psychogenic condition and starts looking for the biological causes, complex as they may be, will physicians be able to relieve the pain and suffering of MCS patients, who contrary to the position expressed in the above passage, want nothing more than to regain their health and get on with their lives.

And it is not just their health that is at risk, according to a February 1998 *Scientific American* article titled "Everyday Exposure to Toxic Pollutants," in which authors Wayne R. Ott and John W. Roberts describe the results of research designed to assess the exposure to toxic substances that people encountered in their day-to-day lives. According to Ott and Roberts, this research showed "results that were disturbing: most citizens were very likely to have the greatest contact with potentially toxic pollutants not outside but inside the places they usually consider to be essentially unpolluted, such as homes, offices and automobiles." [9]

Ott and Roberts concluded that pollution from industrial sources that environmental laws seek to regulate, such as Superfund sites and industrial plants, "was negligible in comparison" to the levels of toxic pollutants inside the home or office. In Bayonne and Elizabeth, New Jersey, cities that have many chemical processing plants, the levels of 11 volatile organic compounds turned out to be much higher indoors than outside. And what were the main sources for poor indoor air quality? Consumer products like air fresheners, moth balls, paint, and particleboard, along with fumes from heating and cooking.

Ott and Roberts note that in the late 1980s, studies of indoor air quality in the cities of Jacksonville, Florida, and Springfield, Massachusetts, found that indoor air contained far higher concentrations of pesticides (typically tenfold higher) than did outside air. To compound the danger, the residues measured indoors included some pesticides that had only been approved for use outdoors. Powerful chemicals used around foundations of houses to eradicate termites were either being tracked into the houses on people's shoes or were seeping through the foundations. To compound the risk, pesticides inside the home are not readily broken down by sunlight and bacteria and "may last for years in carpets."

According to Ott and Roberts, DDT is one pesticide that has persisted in indoor environments for years, even though in 1972 the U.S. government made it illegal for use in this country. Ott and Roberts cite a study by Jonathan D. Buckley of the University of Southern California and David E.

---

[9] Wayne R. Ott and John W. Roberts, *Scientific American*, February 1998.
    <<http://www.sciam.com/1998/0298issue/0298ott.html >>

Camann of the Southwest Research Institute, who found DDT in the carpets of 90 of the 362 homes in the Midwest that they investigated in 1992-93. In fact, Buckley and Camann found that in over half the homes they studied, the levels of seven toxic chemicals were "above the levels that would trigger a formal risk assessment for residential soil at a Superfund site."

As you begin to read these personal histories, remember that the people who wrote them may well be the canaries in the mine who alert all of us to the dangers of toxic chemicals not only in our outdoor environment but also in our homes, schools, and workplaces.

Michael

Pesticide Service Owner

I started in the tree business in southern Connecticut in 1980, started out real small, just a pickup truck, a ladder, and a saw, and I eventually built up to lots of employees, lots of trucks. As my tree company grew, my client base became elite. These folks wanted every square inch of their properties treated for every insect or disease problem. The chemical portion of the business grew. I remember being reluctant to grow this part of the business, but in a service business, the clients dictate. I was doing a half million dollars a year in business, and in a 90-day period in the spring, we would spray over 100,000 gallons of pesticide. The pesticides were not supposed to freeze, so I used to keep them in the basement under my house. It seemed like a good idea at the time, but it doesn't seem like a very good idea anymore.

The pesticide trucks each had a big tank on the back that had impellers in it—things to agitate the liquid inside and move it around. When something would break in there, we would empty the tank, pop the lid open, and jump right down inside the tank and fix it. I never gave it a thought at the time that it was dangerous. I never thought about air quality or anything else inside there. Don't think that was too smart either.

I don't know exactly when I became sensitive, but I do know that it happened over a period of time. When I finally realized that I was becoming sensitive, the first thing I noticed was that I was having trouble with foods. I was feeling sick every time I ate, almost every kind of food. I remember saying that even some water seemed to make me sick, and that seemed preposterous. And I didn't make the connection that I was having trouble with the pesticides until I noticed I would get sick being around the trucks that had the pesticides in them. That didn't make much sense to me, but I started to avoid being around the trucks. I couldn't concentrate, and my short-term memory was poor. I became accident prone and took several falls, so I started shying away from climbing trees and operating our big wood chipper.

One day I went to do an estimate for a well-respected ENT doctor, and he insisted I come into his house after he signed my contract. "I have to look up your nose," he said as he went for his black bag. "How long have you been like this?" he asked. Like what? My nose was clogged, but it had been for most of the last year. I was getting used to it. After looking up my nose, the doctor insisted I come to his office before it opened Monday morning; he seemed very concerned. The antibiotic treatment he prescribed for about three months did nothing, however, but severely increase the gastrointestinal symptoms that had been bothering me. Then I went to the infectious disease specialist at a prestigious hospital. Thirty days of intravenous antibiotics as an outpatient. No change. Sinus surgery, two and a half hours worth, with a severe case of apnea at the end of it. I had to be chemically stimulated to restart my breathing. The surgeon described my apnea as "unusual." The swelling wouldn't go down and the "infection" didn't seem to clear, so steroids and another thirty days of intravenous antibiotics were prescribed. By this time, I was having allergic reactions to the antibiotic, so an antihistamine was added to my IV and two nurses were assigned to watch me, just in case.

The antibiotic treatment didn't help, but I had to keep my business going. On June 6, 1994, all three of my spray trucks were on the grounds of the local hospital and cancer center at 5:30 A.M. to spray pesticide on every tree and bush before the patients and day workers arrived. I had avoided this part of the business all spring, but this job needed my personal supervision. I stayed back from the action in my truck with the windows up. Suddenly I saw a nurse walking straight into the spray area. No one else had seen her, so I jumped out of my truck and intercepted her. I climbed back into the truck and was sipping coffee when I started to feel funny. As I picked up the cup, I noticed a fiery red rash on the back of my hand, then I saw it on both hands, then on my face and neck. I was quickly overcome with severe flu-type symptoms. Although I was within walking distance of the emergency room and doctors, I never thought there was any need to go inside. I just wanted to go home. One of my employees had to drive me home. I curled up in a ball for three days, much more interested in survival than causation.

From that point on, I was unable to work in my pesticide business. I seemed to be reacting to everything in my "normal" environment. Every time I drove my truck I felt ill, and my reactions caused such mental confusion that I had to pull over, lost in places in my hometown where I had been hundreds of times. I really thought I had lost my marbles.

One of the things that happened after I got sick was that I developed such extreme fatigue that I was basically unable to stay awake. I used to sleep 14, 16, or more hours a day. I couldn't stay awake more than 3 or 4 hours at a time, and that phase lasted for easily over 6 months. I can remember getting up in the morning, sending the guys out to work, coming home at 10:30 in the morning, falling asleep and staying asleep till the middle of the afternoon, waking up when the men would come back in, putting the trucks away, and going back to sleep at 6:30, right after dinner. I would sleep 12 hours and get up and still be exhausted, like there was no amount of sleep that would cover it.

People—my wife, my family, other people, my brother especially—used to complain about what a workaholic I was, how I was just unbelievably driven, I would never stop. I would work weekends, I would work till dark. I just couldn't get enough. I was really, really driven to work. So it wasn't like I was really looking for this to happen. It really ruined my workaholic ethic because you can't be a workaholic and be asleep at the same time.

Going from being a workaholic to wondering how you're going to provide for your wife and your two children is pretty tough to take, especially as a male in this society. Society expects you to provide for your family, and people look at you strangely when you don't seem to be working your normal 40 hours or more. It's hard for my wife to understand how she married a guy who at this age is looked at very strangely by the neighbors and a lot of family members because nobody really understands what's going on with me besides the people in my house.

I've been a tremendous burden to my wife. I know that, and at times I've been extremely frightened that she would just wake up one day and say: "I've had enough of this crap. I'm out of here." She could get a better deal almost anywhere else as near as I could tell. That's always been a concern. I've had to live with that, and when I'm really sick and I'm down and out, boy, I can get myself into a wicked hole where all I'm doing is lying in bed and thinking maybe I'll get up and go downstairs and she's finally had enough and she's gone. I hear the car go out and she may have only gone to the grocery store, but I think maybe she's had enough, she's not coming back. I wouldn't blame her.

It's also hard for both my little sons because all the other kids' fathers, they go to work every day. They're not hanging around a lot, and I've had to answer some difficult questions. My sons try to understand, but it's very difficult to explain to a five year old or eight year old why Daddy doesn't do the same things as all the other kids' daddies. Why Daddy can't go out to the soccer practice at sundown when everyone is covered with bug spray or go to the beach when they all are covered with suntan lotion or any other number of things that have happened. But my sons have come to understand it because every person in my family has seen me have a major reaction to something that was very innocuous to everyone else who was at one of these events.

I come from a big, tight-knit Irish family. I don't know if they get any tighter. I've been sick a long time now, and as much as I've explained this to everyone and told them and Judy has told them, relatives still come to visit and wear perfume. We go to visit my mother, and she's got air fresheners hanging all over the house. She's painting with stain-killing paint while we're visiting there. She'll say, "Well, it's in the other end of the house. You don't have to worry about it." People just don't understand how sensitive you can be and how sick you can become from household cleaning products or their perfume or their aftershave or little things like that. I cannot get in my mother's car. I cannot ride in her car because it just smells so much of perfume. Even if she doesn't have any new perfume on, her car just smells of perfume, and I can't even sit in it.

———————————

When you're very sick, you don't have much left, but I've never lost my sense of humor. I've tried to hang on to that because I think if I really did lose my sense of humor, I would start crying and I would never be able to stop. The other thing is, I really think having a strong faith has been extremely important to me in hanging on, not just through this illness but throughout my whole life, through all kinds of things. It's always been there for me. I grew up having that, I was brought up that way. I've got to admit that I've had to test it really hard a couple of times. I don't know how anyone makes it through anything, much less a long-term chronic illness or any kind of chronic life situation, without having something that they can really hang onto. You can call it whatever you want, but I think it's really important, and I wouldn't have made it without it.

Judy and I have always gone to church. In fact, on our first date we went to church before we had dinner. We used to live not very far down the street from this little church. We got married in that little church, and we had both our children baptized in that church. When I started getting sick, before I realized what was going on, I noticed that when I went to church I would be really ill by the time the service was over, if I could even make it through the service. Sometimes I literally had to stumble out of church because I felt so ill. I thought I had done something bad and I was just paranoid about sin or something and God was throwing me out of church. I really was concerned about what was going on. I had no clue what the problem was. Now I miss going to church so much, and every once in a while I'll try it, like just yesterday, for example, I tried going to church. I did manage to make it through the service, but I was really sick for about three or four hours afterwards. I was fine when I got there, but about halfway through the service some woman came in late and sat down a couple of rows in front of us. The eau de whatever she had on—she smelled beautiful, but I could not handle it. I managed to stay through the service, but I was extremely mentally confused by the time I got out of there and just real foggy and it didn't clear until after lunch. It's upsetting not to be able to go to church and not to be able to go to other functions, but church is especially bad because everybody gets all gussied up to go in there.

I still have to be really careful because all I need is a slight exposure and I'm in trouble. For some reason, petroleum will really set me off. Even though I love my tractor, there are quite a few days when I really can't use it. The fumes are just too much for me, I can't get near it. I usually have to have someone else get the fuel for it, and even put the fuel in. Then if any fuel spills, I have to wait for it to dry off completely before I get on the tractor because the smell of diesel fuel, just about any kind of petroleum, can make me real sick, and it makes me very angry sometimes. I get really agitated if I go to the gas station and fill up my own gas tank. Lots of times I get very agitated from the fumes, and I feel like ripping the jugular out of the gas station attendant.

As for the past, when I made my living spraying pesticides, I cringe every time I think of the people whom I mishandled when they asked us to notify them when we sprayed their neighbors' property or harassed us while we were legally filling our trucks at streams with a suction hose. (We were accused of dumping, not filling.) Both sides approached the issues rigidly, resulting in misunderstanding all around. The biggest curse of all though is the harm that is being done to the applicators of pesticides. More than once I let go employees who complained of symptoms that they associated with applying chemicals. I thought they were just malcontents with imaginary complaints. The warning labels on the pesticide containers are incomplete, and there is nothing to keep anyone from mixing chemicals accidentally or on purpose in an application process that could create a deadly toxic cocktail. More and more chemicals are sprayed on residential properties every year.

I can remember a specific incident involving a little cul-de-sac with five houses on it. We sprayed four of those five houses. At the only house we didn't spray, this woman came out in her driveway and literally just started screaming at me about how she didn't want pesticides sprayed in the neighborhood. She obviously didn't know I owned the business. I gave her a serious rash of crap because she had given it to me, and I have thought back about that many times. She was complaining to me that her child was really allergic to pesticides and she was just trying to protect her child, and at the time all I was thinking about was the 450 bucks she was trying to keep me from making on this street. I thought I was just down there like the Good Humor man, I was just down there delivering the ice cream. I didn't think I was doing anything more serious than that. I wish I could apologize to that lady. I'm sure that her little kid probably *was* sick from pesticides.

We moved from Connecticut to northern New England on January 29, 1995, leaving at 3:30 A.M. to avoid traffic and traffic fumes. I drove a U-Haul truck for six hours with the windows down and no heat in frigid weather in order to avoid the fumes from the heater. We moved to get to a cleaner environment, which seems to have worked, although it's taken years for me to notice a substantial improvement. We've been here for almost five years now. I've really started trying to work in the last year. The only other thing I know besides tree work is remodeling houses and working on houses. Since we moved to northern New England, we've worked on three different houses, trying to make them safe for me to live in. We've also been trying to invest a little bit and fix up some other houses and hire people to work on them using safe paints and building materials that don't bother me too much. Then we turn them over and sell them and make some money, and that seems to be working on a small scale for us at this time. That's basically how we're surviving, but it takes both of us to do it. On days that I can't function, Judy ends up running the family, running the business, doing the whole thing. I couldn't do it by myself, that's for sure.

I've been sick over eight years, and that makes it seem very difficult to plan anything for the future although I always used to be a big long-range planner. My plan now is to make it through to the end of the day, make it through tomorrow. I'm planning to the end of the week. It's very hard to plan for long range, when sometimes just surviving minute to minute or making it through the next hour really seems like where you've got to put all your energy. You don't have a whole lot of energy to plan long-term, future-type thoughts. It's very difficult to just go along like that every day. I kind of feel like I got robbed, and although I'd like to think that there's going to be a bright, rosy future, I know that I have to spend my energy focusing on making it through day to day and getting my family provided for and surviving day to day. I have no idea whether I'm going to make it long-term.

Judy, Michael's Wife

After we renovated our farmhouse with all nontoxic substances, we moved in in a flurry pace. Within 48 hours, however, Michael was sleeping outside on the porch under five wool blankets and three sleeping bags because he was unable to live in the house. It was November, and it was beginning to snow, so he was really freezing out there.   We had been renovating the house for four months, and Michael had been in the house every single day with no problem until the last two or three weeks, when we were working on the kitchen floor. We were having it laid with linoleum, and Michael noticed he was losing his voice and not feeling very good. We are convinced that the glue that was holding the linoleum to the floor was what pushed Michael out the door. He was unable to enter the house for another month after that except for very short periods of time. When he did, you could see a red rash develop on his face and he would get beady eyes, glassy, watery eyes.

During the time when Michael was sleeping on the porch, I ran into a friend from high school who lives in this town we had recently moved to. So this guy whom I hadn't seen in twenty years or so said, "Well, let's get together. Come on over to my house." And I had to politely say, "I'm sorry, we can't go to your house." To which he replied, "Well, what's the deal, you can't come to my house?" Well, what am I supposed to say to him? The next time I see him, two months later, he says, "Well, what's going on? Why can't we get together?" So I tried to explain: "Well, my husband has this illness, multiple chemical sensitivity, and he can't come to your house, and people can't come into our house if they're wearing perfume or have shampooed their hair with a fragranced shampoo." And all this year Michael's health has been up and down, up and down, and the guy, every time I see him, he says: "Well, when am I going to get to meet your husband?" I feel like he's a ghost, he's not really there.

It's very difficult to understand, very difficult to explain to people. It's difficult for me, and I've been married to the man for nine years. I see reactions happening with my own eyes, and it's frustrating. It's frustrating to say, "Dammit, I got a sniff of perfume, or there were pesticides on the peach I ate." And then he's down, he's down and he's out, and that's frustrating. How do you live like that? That's what it is. Totally, unadulterated, flat-on-your-back illness. It's crazy to live with that, but you have to live with it, I mean he's my husband. We've made a commitment to each other. The more I work with him, the better and the quicker he will get well again so he can function on a day-to-day basis.

I had to ask this man who works very closely with us who wears aftershave or cologne to please not wear the cologne because every time Michael gets next to me after he's been around someone who is fragranced, he becomes extremely irritable and can have starbursts of anger. For a long time I didn't understand what was happening, but now I know not to take that personally. I realize that when he smells something fragranced on other people, he gets extremely irritable very quickly, and I have to say, "That's what's happening, so you don't need to yell at me."

Michael's update, October 1999: Looking back with a clearer head, I believe my chemical sensitivity started before my active pesticide use but was greatly exacerbated by it. When I was a child, my friends and I used to ride our bikes after the trucks spraying DDT in our neighborhood. We thought it was fun to dart in and out of the mist coming off the truck. Perhaps that was the origin of my chemical sensitivity, but that's hard to determine.

The good news is that after more than ten years wrestling this MCS, it seems to be clearing in the same way it started, slowly, almost immeasurably until a milestone is passed. I can now visit my children's school, ride in most cars, and go in some stores. I also have the energy to play with my sons. As of this writing, I am attempting full-time work (a sales job that I can do using my home as a base) for the first time in almost six years. My little victories are unseen by most, but the cumulative effect is that many symptoms have improved. I tread very carefully, however, around anything that might set me back. When the setbacks come, I rest, regroup, and get off the canvas. Faith and determination are my allies, clean air and food have helped tremendously.

Terry

Gulf War Veteran

I was in the Marine Corps from 1960 to 1966. When I left the Corps, I got a job with a major railroad company as a steel worker building rail cars. I started a family and earned a college degree by attending classes at night while I worked for the railroad.

In November 1990, I was activated with my Army National Guard unit and sent to the Persian Gulf. I served both as a medic and as an intelligence NCO. Prior to the Gulf War, I was in perfect health and in great physical condition. Now I'm in a wheelchair.

I spent that Christmas Eve of 1990 in an unforgettable sandstorm that some said was one of the worst sandstorms that had ever been recorded in Saudi Arabia. The storm hit on Christmas Eve, when we had just been trying to put together a makeshift Christmas tree using chemical lights— those plastic straws that glow in the dark. Anyway, the wind quickly became so strong that everyone inside our tent was ordered to hold the sides of the tent so it wouldn't blow away. The strength of the wind blew fine sand right through the canvas, so we had to put our ponchos in front of our faces to keep from being choked by the dust. It was a sandstorm and a Christmas Eve that none of us who had arrived in the Persian Gulf early will ever forget.

One of my jobs as a medic was to pass out malaria pills to the troops in our unit every day. There were lots of mosquitoes around, so malaria was a real threat. As a result, we used a lot of DEET. We had two kinds of government issue—a cream to put on our skin and spray cans containing what you might call industrial strength DEET. We weren't supposed to spray the stronger version in confined spaces, but lots of soldiers used it when they ran out of the cream, which was pretty often. They would spray it on their skin and uniforms while they were inside the tent. Some troops even wore flea collars around their wrists and ankles.

We were stationed in Riyadh in Saudi Arabia for much of the Gulf War. There were lots of scud attacks on Riyadh. Almost every time there was a scud attack, our chemical alarms would go off, but higher authorities claimed the instruments were all defective and were giving us false alarms.

Like almost everyone else in the Gulf War, I had to take the pyridostigmine bromide pills. About three days after I started taking them, I developed a stomach ache and nausea. I saw some other people throw up and even saw some people throw up blood. Women had a worse reaction than men. The pills didn't seem to bother some people at all, however.

When we returned to the United States, we received the most minimal physical you could ever give to servicemen returning from war. I went back to work with the railroad but continued to serve in the Army National Guard. After returning to work, I started having problems that I had never had before. The automatic air freshener dispenser in our office was making me sick, as were paint and diesel fumes. I started having panic attacks, and I was having problems concentrating on my work and other activities. I no longer could climb to the top of railroad cars. One morning I had a terrible headache, a stiff neck, and a very high fever. I was rushed to the emergency room, where they did a spinal tap. The doctor did not find anything, but he treated me for spinal meningitis.

Everything seemed to get worse from that point on. I was falling down at work for no good reason. In January 1993, I could not pass the Army physical fitness test and had to retire from the Army National Guard. That year I started using a cane because my right ankle and leg just kept giving way as I walked.

After I left the National Guard, I went to the VA hospital in my area and signed up for the Gulf War Registry and took the Gulf War veterans' physical. I was placed in the hospital for one week. My condition got worse, and in March 1993, I was sent to the VA hospital in Washington, D.C., for more tests. I was only supposed to stay in the hospital for seven to ten days, but they kept me there for six weeks.

My discharge summary was panic disorder, degenerative joint disease, severe high frequency hearing loss in right ear, allergic rhinitis, emphysema, high cholesterol, chronic fatigue syndrome, possible multiple chemical sensitivity, mild patchy sensory neuropathy of distal nerve segments, and multifocal motor neuropathy with conduction blocks of the ulnar and radial nerves of both arms. The doctors stated regarding all the above medical problems: "etiology unknown." The doctor who told me I had multiple chemical sensitivity said he wasn't allowed to write that in the diagnosis—he could only say "possible multiple chemical sensitivity."

In 1994, I had to start using two canes, and that worked for about six months. If I happened to smell something like gasoline or perfume, however, it would affect my coordination and I would fall down. Finally, the Veterans Administration gave me a wheelchair, and a couple of years ago they even gave me an electric one because I have a lot of nerve damage in my arms. I can't complain about the medical care the VA has given me. They have paid for my medications and hospitalizations, even though they claim my problems are not service connected.

When I returned home from the VA hospital in Washington, the railroad company would not let me return to work and retired me on disability. I could not understand what was going on because I had always been very healthy before I went to the Gulf War. I contacted my VA representative to file a claim for service-connected disability, but I have not received anything. Other vets who worked right beside me in the Gulf War have received disability compensation for diagnoses like fibromyalgia and chronic fatigue syndrome, but I don't know of anyone who has been given a disability award for MCS.

When I went to a civilian doctor in connection with my application to obtain Medicare coverage from the Social Security Administration, I happened to mention that I had MCS. He went ballistic and said, "So you're one of those people. Let me tell you what, you just lost all your credibility with me." He turned in a negative report to the Social Security board, which then denied me Medicare coverage. I have, however, been on Social Security disability payments since 1995.

I have had several other medical problems in the past eight years. I have skin rashes and sores, stomach problems, and chronic pain. Before the Gulf War, I would rarely take an aspirin or any other kind of medication. I worked very hard all the time and never took time off from work because of illness. Now I take 25 pills a day. I can no longer drive, and I am having a very hard time adjusting to that.

Before the war, I had been stung by bees on several occasions and never had a bad reaction. Now bee stings bring on anaphylactic shock. One time when I was stung and my wife was rushing me to the hospital, our route was blocked by a passing train. She quickly whipped the car around and tried another route, only to find that a truck full of watermelons had overturned and blocked that road. She was in a panic by that time because my swelling was increasing and my throat was closing off so that I could barely breathe. Fortunately, by the time we returned to the first road, the train had passed. When we got to the emergency room, I was in such bad shape that a nurse came out to the parking lot and had me drop my pants right there so she could give me a shot.

My problems with chemical sensitivity began right after the Gulf War. I still cannot pump gas or diesel fuel; my wife pumps it while I sit in the car or truck with the windows rolled up. The ink smell on a fresh newspaper really bothers me. I have to watch what type of shampoo I use or I will break out in a bad rash. Some perfumes also give me a bad reaction. Just last week I started to read a magazine, and it turned out it had a perfume strip in it. I got nauseated immediately and got a headache; I felt so sick I had to go to bed. The next morning I was still nauseated and was wheezing, so I had to go to the hospital. They kept me there a couple of days. I had another bad reaction to perfume when I went to my grandson's kindergarten graduation. I had to be carried out of the building because someone was wearing a perfume that was so strong that I could not breathe and got very sick.

It is very hard living with multiple chemical sensitivity because there are so many chemicals that can make you really sick very fast. To me MCS is not a joke and should not be taken lightly. It's had a terrible effect on my life.

Editor's note: On October 26, 1999, Claudia S. Miller, M.D., M.S., stated in testimony before the Committee on Veterans' Affairs, Subcommittee on Benefits, U.S. House of Representatives, "Over the past several years, the finger has been pointed at a number of potential causes for Gulf War syndrome—everything from the oil shroud to pesticides, vaccinations, and pyridostigmine bromide. What set off the Gulf War veterans? The answer is 'all of the above.' Exposure to any one or any combination of these toxicants may, in fact, be capable of causing a general breakdown in tolerance that can result in a plethora of beguiling symptoms." For the complete text of Dr. Miller's testimony relating Gulf War syndrome to chemical sensitivity or what she terms Toxicant-Induced Loss of Tolerance (TILT), see Appendix 3.

Erica

Physician

I was in very good health before I got exposed to a number of different chemicals in my workplace in a clinic. Before that happened, I was extremely active outdoors, a world-class athlete. I did a number of things, but most significantly, I led international expeditions around the world.

After working as a schoolteacher on a Navajo reservation for a few years, I went to join the Peace Corps and worked in South America. At that time I did a lot of high altitude climbing, and I was the first American woman to climb Aconcagua, which is the highest mountain in the Western Hemisphere, in Argentina. I also did a first ascent of a 17,000-foot peak, which was subsequently named Pico Erica. That is the custom down there, to christen a peak that has never been climbed before after the person who first climbs it.

In 1980, after my first year in medical school, I led an all-women's expedition to the top of Mt. McKinley, also known as Denali, in Alaska, which is the highest peak in North America at 20,000 feet.

Since becoming chemically sensitive, I found that for the first few years I had such neurological damage that I was no longer able to do the kind of sports that were my passion. I could no longer rock climb, no longer ski except on the baby slopes. I found myself falling down, tripping over my own feet, tumbling downstairs. But gradually over the years I've been regaining my neurological function so that now I am able to run again, which is a godsend for me. It was a terrible loss when I could no longer run.

My health problems began in 1991, when I went to work for a large multispeciality clinic. I was exposed to many different chemicals in a building that was subsequently designated a sick building. Several other employees were also affected by the chemical soup there. I suddenly became extremely tired. Normally I'm an exuberant, high-energy person, as all my acquaintances can attest, but I started having a difficult time getting out of bed in the morning. I felt like I could fall asleep anytime, and I was having trouble keeping my head up while I was talking to patients. I also had difficulty following what they were saying. I noticed, however, that on the weekends I felt better.

I started getting quite a bit worse. I began getting rashes and joint pain, and eventually I no longer felt well on the weekends. I felt bad all the time. And also I noticed that periodically I would get these flu-like symptoms, and one day I got this terrible chest tightness and shortness of breath. When I asked if anybody had used any different kind of chemical, I found out that an organophosphate pesticide had been applied that is known to be neurotoxic. When I called the company that applied the pesticide, I found out that their monthly applications coincided identically to my journal entries noting when I had had flu-like symptoms. Gradually I began putting the pieces of the puzzle together, and that's when my extensive research began.

There was also a new carpet in the area where I worked that was just a year old and was offgassing a lot of chemicals. There were no operable windows in my area. I also noticed a very foul smell, and when the environmental engineers came, they discovered that the HVAC (heating, ventilation, air conditioning) system was not hooked up correctly. The exhaust fumes were funneled right into the air intake when the wind blew from the west. As it turned out, the outlet over my head at the nurses' station was positioned in the system in such a way that the exhaust fumes hadn't had much chance to be diluted before they came out into our air space, so my nurse and I were very heavily affected by these exhaust fumes.

There were other flaws, and altogether they contributed to the chemical soup in that clinic. It was not just one agent; there were many agents. Disinfectants were a major one, however, and pesticides were also a big contributing factor.

Fortunately, my health has gradually improved over the last few years through avoidance of chemical exposures. In 1993, I had a house built for myself with nontoxic materials in a development that does not allow pesticide use. Having a nontoxic home has been the factor that has contributed most to the improvement in my health. I also changed my diet drastically. When I eliminated dairy products from my diet, my fibromyalgia disappeared. Eliminating processed foods improved my intestinal problems. I now eat organic foods as much as possible and eat a diet that includes meat but is heavy in vegetables.

I have recovered my health sufficiently to be able to practice medicine in an office in my home, where I primarily treat patients with multiple chemical sensitivity, chronic fatigue immune dysfunction syndrome, fibromyalgia, auto-immune diseases, allergies, and other conditions related to immune dysfunction.

Pat

Gulf War Veteran

I was in a transportation company that went to the Gulf War. My health was good, and I had no symptoms before the war. For the previous six years, I had driven a tractor-trailer truck and had been around diesel fuel and exhaust every day without it causing any problems.

My unit arrived in the Gulf the day before the air war started. We first spent about a month in Dhahran in Saudi Arabia. Our chemical alarms went off several times during that month, and we had to go to MOP-level four, which meant we had to put on chemical suits, masks, gloves, and boots. While we were still in Dhahran, we started taking pyridostigmine bromide pills, which were supposed to protect us against exposure to nerve gas. About three days after I started taking the pills, my eyes were jittery, my vision was jumping, I was seeing double, and I was nauseated. By the fourth day, I was vomiting a little blood, so I went to sick-call. They told me to cut the dose in half and said there was nothing to worry about. At least I no longer vomited any blood after I reduced the dosage. Many other people in the unit reported having similar vision problems.

After we left Dhahran, we moved up north. When we got to our encampment, the first thing we had to do was to set up all our tents. Pesticide was sprayed around the periphery of each tent, and later on we used spray cans of pesticide inside our tents to keep away scorpions, sand vipers, fleas, sand ticks, and flies.

About a week after we moved into this encampment, the authorities decided to spray diesel fuel on the ground all around the camp to keep the sand down because trucks would stir up the sand and dust and sometimes even get mired down in it. The blowing sand was a problem everywhere we went in the desert. When we drove our trucks in convoys, we had to wear bandannas over our face to protect ourselves from the sand kicked up by the truck ahead of us. Even then we were eating sand every day.

The water we bathed in and washed our clothes in was oily, so after a shower we would have an oily film on our skin that smelled like diesel fuel. We later realized that the same trucks were often used to carry water and diesel fuel.

When I was in the Persian Gulf, I drove over a big pothole one day and hit my chin on the steering wheel. The impact broke off two crowns. By the time I got back to the States, all my teeth were beginning to deteriorate and feel loose, but I don't think that was related to the pothole incident. I ended up having to get dentures at age 35.

When I came back from the Gulf War, I was in bad shape. The Veterans Administration in North Carolina gave me a wheelchair because I could barely walk. I had to use the wheelchair for about a year, and then I used a cane for a year. At any rate, when I got back to the States after the war, I was very, very weak and I had swollen and painful lymph nodes in my legs, my groin, and my neck. I was eventually sent to the Veterans Administration in Washington, D.C., for evaluation. The VA just didn't get it about the connection between MCS and Gulf War syndrome. When they were treating us in the VA hospital in Washington, they were painting the hospital wards, and it made all of us with Gulf War syndrome deathly sick. We told them and they wrote it down, but that was all that happened.

While I was in the VA hospital, a surgeon did a lymph node biopsy of my left cervical and accidently cut the spinal accessory nerve, so now I have spinal accessory nerve palsy. Due to this nerve damage, the muscle has wasted away on one side. My left shoulder now droops four inches below my right one. I can't shrug my left shoulder, and it's hard for me to raise my upper left arm. I also have spinal problems now because I can't stand up straight.

At first after the operation I thought I was just having pain because of the surgery. I didn't realize that the VA in Washington had botched the operation until I was back in Mississippi. I begged the VA to pay for me to travel back to Washington to have them see what was wrong, but they kept refusing to do so. It was a month before I could get permission to go back to Washington; I finally had to get my senator and congressman to pressure the VA. One of the reasons I was fighting so desperately to get back to Washington was that I was afraid that the nerve damage was going to cause me medical problems for the rest of my life. I knew I would have to pay my own bills for treatment unless I could establish that the injury was service connected because it happened in a VA hospital.

While I was back in Washington, they finally sent me over to Bethesda to see a neurologist there. He just shook his head and couldn't believe what had happened. Back at the VA hospital, one official said to me as I was crying about the damage to my shoulder: "Oh, Ms. B., that's not really a problem. You can wear a shoulder pad. All the ladies' clothes have shoulder pads in them." I won't tell you what I told him, but he got out of the room real fast, and I got my letter the next day saying that the injury was service connected.

I was a reservist for a number of years after the Gulf War, but even when I didn't need a cane anymore, I couldn't drive a truck because every time they would crank up the trucks, I would get deathly ill from the diesel exhaust. The only thing I could do was office work. Eventually they put me on the temporary retired list. After I've been on that list another year, I'll be put on permanent retirement.

When I first came back from the war, I had a lot of problems remembering things. One day I drove my kids to school and dropped them off, but then instead of driving home I just wandered around for a couple of hours because I couldn't remember where I was going.

I had real weak muscles when I returned from the Persian Gulf, and I also had sore places on my hips, like fibromyalgia. I have migraines now, and I didn't have them before the war. After I came back, I had two duodenal ulcers, and I've developed gastric reflux. Every day I run a low-grade fever (99.8-100.3), just enough to make me always feel a little bit yucky. I have sinusitis all the time now and take antibiotics for that. I have also tested positive for *mycoplasma incognitus (fermentans)* like so many others with Gulf War syndrome.[1]

Since I returned from the Gulf War, I can't be around paint fumes because they make me really sick. Bleach also bothers me now, and I can no longer wear perfume. My life has been changed in so many ways by my experience in the war, and chemical sensitivity has been a big factor in those changes.

The Veterans Administration says I have panic disorder. The military says I have PTSD—post traumatic stress disorder—but the VA says women can't get PTSD. I guess we can drive tractor-trailer trucks, but we can't get PTSD. After I got back to the States, I did start getting panic attacks when I drive, so I no longer drive much.[2] At any rate, there's no way you can write off everything that's gone wrong with my health by calling it panic disorder

---

[1] Many Gulf War veterans and members of their families have tested positive for *mycoplasma incognitus (fermentans)*, a bacteria without cell walls that Garth Nicolson, Ph.D., believes is the cause of Gulf War syndrome. Clinical trials of antibiotic therapy are now being carried out by the Veterans Administration. See Appendix 2 for a listing of an article on *mycoplasma incognitus (fermentans)*.

[2] Many people with MCS report having panic attacks after exposure to certain chemicals. Pat is now extremely sensitive to diesel fuel and exhaust, so it is possible that her panic attacks while driving are related to that sensitivity.

or PTSD. The VA just needs to get it together and realize that people are actually dying out there or are in misery or pain, and they need help.

I'm not well, but I am better. Maybe you just come to realize that this is the way it's going to be and you just accept it. I don't dwell on it anymore.

Jeff

Industrial Plant Worker

My problems began from working in an eight-foot deep concrete pit at the company where I had been employed for four years. My company was a subsidiary of one of the largest corporations in the country. Both my brother and I got the worst of the exposure. One other person got exposed, but to a lesser degree. The pit contained a large milling machine we were leveling, and there was another little pit in the corner of the main pit that was filled with coolant. My brother and I were the first to go down in the main pit to prepare the machine bed for leveling. After about 15 minutes, we both felt like we were drunk, and we reported this to our foreman. He reported it to Bob Mason (not his real name) in the health and safety office, who informed our foreman there was nothing down there and told us to keep working. We set up fans and blowers to try to air out the pit.

We worked in the pit for several weeks, getting sicker and sicker all the time. Some of our symptoms were headaches, dizziness, drowsiness, confusion, anger, skin rashes, ringing ears, pain in our joints, twitching eyes, and vision problems. We finally talked Bob Mason into providing respirators for us. He wouldn't provide them "on the record," however. He said he would leave them on the table and we could take them. At first the respirators seemed to help us spend longer periods in the pit, but after a few days, they seemed to lose their effectiveness.

At this point we were talking about quitting our jobs, so another employee, who was not from the health and safety office, decided to do a cleanup of the pit. The first thing that was done was to suck out the coolant pit with a vacuum truck. The truck operator never even went into the pit. During this operation, I went over to the edge of the pit for five minutes and got extremely sick. The guy operating the vacuum truck said the coolant was hot. I felt the hose to confirm it. He said it may have been a chemical reaction or a faulty circulating pump. I noticed the truck operator was absent the next day. A maintenance man worked on the leaking pipes of the suck truck the very next day, and he got very sick. The men in the crew that cleaned up the pit were also telling us they felt sick. While the pit was getting cleaned up, my brother and I were working in another area. We noticed our symptoms were getting worse anytime we came into contact with any cleaners or oils, things that had never given us a problem before.

After the cleanup, my brother and I went back to the pit, and we soon started deteriorating even further because there were still chemical fumes in the pit. At this point we said enough is enough and packed up our tools. Our action caused Bob Mason to finally come down from the health and safety office (for the first time, I might add) to see what was going on. He decided to get some samples and actually got me to go into the pit to get them. We left work in hopes that our health would return, but it never did and it's been over five years since that exposure. Both my brother and I have been diagnosed with MCS.

I tried to see a doctor who is an industrial hygienist at the medical school of our state university. He asked me to get a list of the chemicals I had been exposed to so that I could bring it to my first appointment. I finally got Bob Mason to give me a list of the chemicals I was exposed to, and the list included 1,1,1-trichloroethylene, toluene, butylated hydrotoluene, and traces of methylene chloride. This list, however, was based on tests of chemical levels that were performed after the coolant had been sucked out of the pit. I've never been able to find out what chemicals were in the coolant because that's a trade secret. The health and safety office always claimed that there had been no tests done before the coolant was sucked out. I knew differently because one day prior to the cleanup I had seen a man taking samples in the pit. I went out to the plant parking lot and saw a van clearly marked with the name of a chemical testing company. Later on, however, the health and safety office denied that samples had been taken before the coolant was pumped out.

At any rate, when I made the phone call to Bob Mason to get the list of chemicals I had been exposed to, I made the mistake of telling him I had an appointment to see a physician at the university medical school. He asked me the physician's name, and he said he knew him. One hour after I finished talking to Bob Mason, the medical school called me and pushed my appointment back another month. That still seems like a strange coincidence.

At this point, I called OSHA, and they recommended a physician by the name of Dr. Smith. I initially saw his assistant, Dr. Walker, who seemed pretty good at recording my symptoms and workplace exposures. When I returned two weeks later, I was still quite ill. I told Dr. Walker about my problems driving, entering most buildings, and encountering substances like diesel, paint, cleaning products, or perfume. I was unable to work in this condition. There were about five doctors in the room on this occasion, including Dr. Smith. Looking back, I wonder if my case wasn't a hot potato because I worked for such a major corporation. Dr. Walker said that my symptoms were consistent with exposure to the chemicals found in the pit, and he said he would support my claim to workers' comp. Dr. Smith was obviously not happy with Dr. Walker's assessment of my case. He gave Dr.

Walker an extremely angry look and said my symptoms should be all cleared up in a few weeks.

I returned to the clinic a month later because my symptoms were still bothering me. At this point I saw only Dr. Smith. He examined me and said I was fine. I was feeling extremely sick from the ride there and from being in the office and was actually shaking quite badly. I said, "What do you mean, I'm fine? Look at me!" At this point he suggested that I see another doctor for neuropsych testing. That doctor said I had MCS.

Nevertheless, workers' comp kept sending me to doctors who would say I was OK. This eventually led to me getting kicked off workers' comp. As it stands now, I am unable to work, and it looks like I may be like this for quite a bit longer, judging from the past five years. It's been a constant battle since I had the exposure in that pit.

I finally got a lawyer to try to help me get workers' comp, but that turned out to be a disaster. During the hearing, I had to watch the judge help the workers' comp lawyer ask this question or that question. My lawyer just sat there like a lump taking it all in while I was getting sicker and sicker on the stand because of chemical exposures in the hearing room. He didn't make objections or do anything to help me win my case. As a result, I got removed from workers' comp.

Later I began to read about my law firm in the newspapers because the state attorney general's office raided the firm and seized their files. One newspaper article said my law firm had been indicted on 82 counts of insurance fraud, motor vehicle insurance fraud, larceny, attempted tax evasion, and conspiracy. My lawyers just told me that the government was out to get them because they were helping injured workers, and unfortunately I accepted their explanation. They said they couldn't handle my case any longer, but they set up a meeting between me and another lawyer who came to their office to talk with me. He agreed to handle my appeal, but after that day I never could reach him. I called repeatedly, left messages on his answering machine, and wrote to him. He never replied. Finally, after about eight months I succeeded in talking with him, and at that point he told me that the statute of limitations on my case had just run out. To this day I still wonder what was going on behind the scenes.

It's really discouraging when TV shows and radio shows and articles try to portray MCS patients as crazy. I know that may affect the way some of my friends view me. I know how silly I look wearing that stupid, uncomfortable mask when I go some place where there are strong chemicals. We need definite tests to shed some light on MCS. I don't like being treated like a nut case when my health has been wrecked by my exposure to those toxic chemicals in the pit.

Marilyn

Clerical Supervisor

In 1994, my co-workers and I were working in a state office building where we were inadvertently exposed for seven months to an unvented printing press. The printing press exposure involved solvents like 1,1,1-trichloroethylene. Then throughout the fall and early winter, pesticides were also sprayed around our desks as we worked. The applicators sprayed the pesticides during our breaks and while we were eating lunch at our desks, as at that time we had no designated break room. The instructions for the use of the particular pesticide that was sprayed clearly state: "Do not remain in treated area and ventilate the area after treatment is completed. . . . Do not contaminate water, food, or foodstuffs." The clincher was that the building, which was already saddled with a history of air quality problems, did not have an operable ventilation system. I would compare the situation with running a car in a garage without the door open. According to statements made by the architects during a meeting in January 1995: "The ventilation system of the entire building was on recirculation mode. This resulted in all irritants in the facility being effectively sealed within the building." Were we lucky or what?

In our area alone there were 20 people who complained about symptoms at work after the printing press was installed, and seven of us ended up with what would appear to be lifetime disability from the exposure. My friend Jeanne and I sat directly over the printing press, which was located on the floor below us, and we are two of the people most affected by the exposure.

When I was hired in July 1992 at age 43 by the state government department where I still work, I was an extremely healthy person. I saw a doctor yearly for those fun paps and breast exams that I had to drag myself to, and that's about it. Then in October 1992, I had the first sinus infection of my life, and I also began to have a lot of fatigue symptoms simultaneously with chronic sinus problems. I jokingly said, "I'm getting old" to other colleagues I worked with, but some people said that I was just joining the rest of the building. I heard many people say that they had had constant bouts of bronchitis, pneumonia, and worsening asthma since working in the building, but at that time, I disregarded my symptoms, thinking they were just medical problems that were unrelated to my workplace environment. We didn't know at the time that the building ventilation system was on recycle.

During July 1994 when the printing press was installed, my friend Jeanne began to break out in a terrible rash. As a child she had been diagnosed with eczema, but she had had no outbreak for 25 years. Now she became covered with this skin eruption and was seriously affected by it. The eczema began on July 12 that summer. (We later learned that they began running the new printing press on that day.) From that day on until the present, she has had to see specialists regularly for this condition. Her whole body was affected, but in particular we saw her face break out, turn purple, and peel in great sheets. It was repulsive, and she felt terrible, but we all loved her anyway. Jeanne went through the summer and into the fall like this, and the eczema got even worse.

In mid-October, when it was beginning to get cold, they shut the windows in the building. I began to have bouts of shortness of breath and needed to leave the building to get enough breath, although I had no history of asthma or allergies. I had never before had a rash in my life, but all of a sudden my face was bright red and burning every day. My eyes would itch, and I had a skin rash across my chest and up my arms. It would go away when I left the building and begin again when I went back in. Polyps began erupting on my face, and my mouth would begin to tingle and then burn. My lips would go to sleep and sometimes swell a little when I was in the building. My tongue and airways burned. There were times when I was not able to concentrate and I would slur my words, my tongue feeling too big for my mouth. At night I had intestinal cramps that were alarmingly strong. I began to keep a daily account of the symptoms because I was aware that something was happening, but I had no idea what. My family began to worry, and rightfully so. It was not Mom as usual.

My symptoms became worse toward winter, and by mid-December 1994, many of my co-workers also were noticing that they didn't feel well in the building and felt much better on weekends or even just when they went outside into the fresh air on a break. People began to talk to one another about the way they were feeling, no one understanding what the exact cause was, but all understanding that it was connected with being at work.

Then one day our director noted a foul odor. He called the head of building operations, who promptly appeared, and he could smell it too. After that, an environmental health firm was called in to investigate. An architectural firm was also called in, and industrial hygienists were eventually hired to do air quality tests. The consultant from the environmental health firm moved Jeanne and me out of the building to another site. Jeanne never worked in the building again. At the time I thought it was awful that she had to leave, but in retrospect, I was the one who was not so lucky.

The group of investigators found several OSHA violations such as chemicals in the break rooms and on the heating vents, the new unvented printing press on the floor below us, the lack of a functioning ventilation system for the building, humidifiers containing bacteria and mold, air vents that had not been cleaned for decades, air diffusers loaded with bacteria at dangerous levels, and other health hazards too numerous to mention. Bad air quality tests had the entire group of investigators calling this a sick building.

Two and a half years later I read the minutes from an air quality meeting of all those involved. The health firm representative said: "Both individuals [meaning me and my friend] interviewed were showing significant symptoms of reaction. . . . There was a high probability that people within the building were experiencing symptoms without speaking up. . . . The evidence to date, the MSDS sheets [material safety data sheets], the nature of the symptoms, the locations all point to a potential level of hazard that is defined in some cases as life threatening." I read this in quiet shock. Even after I had become permanently sick, I had never realized how endangered we all had been at the time.

Meanwhile, I had just been hired to be the clerical supervisor of our legal unit, and I was thrilled about it. I was unable to do my job off-site they said, so back into the building I went, but in another location upstairs.

About two weeks after I had been in the upstairs site, my boss came to me and wanted me to move back to my old work site. He said, "If you can't do this job, we'll find someone who can." I got the message and considered it a threat. I was a single parent of three children, one out of the house, one in college, one still in high school, and me the sole provider living on a shoe-string budget. My boss threatens my job, my very existence, my children's security in one sentence showing who had the power, and let me tell you, it wasn't me.

I sat and stared out the window for a long time after he left. I said things to myself like, "Well, are you going to whine and cry about this or are you going to go do your job?" The debate raged on: "Are you going to get sicker here and ruin your health?" And I imagined those bill collectors saying, "My money is due on this date and that date." It was a horrible decision to have to make, and I didn't know what to do. I finally told myself that I would go out of the building to breathe some fresh air when I needed to, and I would just put up with the burning sensations, rashes, and breathing difficulties as I had been doing for several months anyway. I would deal with problems as they arose. If it had been just me, I would have left in a minute, but my middle daughter deserved to stay in college and my younger daughter was in the middle of her senior year of high school, with its extra expenses. So I went downstairs to my old desk in the unit where the air quality was the very worst in the entire building and sat there for another six months reacting every day.

Then in May the granite steps leading into the building were redone. They were only about 40 feet from my desk. It was incredibly humid and in the nineties, so people began to open the windows along the side of the building facing the construction. Cement mixer dust, gas generator exhaust, and epoxy fumes drifted into the building. I began to react. I sat there for six hours while they worked on those steps. My eyes felt as if a pile of sand had been filtered under my lids. It hurt to move them. My breathing was raspy, and I was losing my voice. I was wiped out with fatigue and was unable to concentrate. Welts appeared on my neck, and my skin was on fire. My face and mouth were asleep and burning so bad I could not touch my tongue to the roof of my mouth without pain. I was sick to my stomach and had a terrible headache that I can still remember, so I finally went to my boss and asked if I could go home. When I told him I was reacting to the construction fumes, he rolled his eyes, but I had to go home sick anyway at 2 P.M.

Unfortunately, the symptoms that usually dissipated by the end of the 15-minute ride home did not disappear, and I returned the next day to the same construction situation not fully recovered and reacting anew. Overnight a group of five polyps running from just under my left eye down my cheek appeared on my face, and my doctor later said they were from "skin trauma." I toughed out another partial day and then made an appointment with my doctor. I didn't realize it at the time, but I had gone over the edge with this second major exposure and my life would never be the same again.

My doctor had known about the solvent exposure from the printing press, and when I went to him after the step renovation exposure, he insisted I leave the building. Enough. I couldn't have stayed there anyway. Even if Christine dropped out of college, Beth didn't even graduate, and the bill collectors camped in tents on my lawn, I couldn't have worked in that building. I was sick there each time I walked in, and I couldn't fight it.

After the exposure from the step renovation, I began to become sick in other places. One day when I went to another building for an appointment, the person helping me said, "You're getting really red and your voice is hoarse. Are you all right?" I replied, "I do feel kind of funny." In about ten minutes, you could hear me breathing, so I left the room and headed to the elevator. It was four floors down to fresh air, and I almost didn't make it. I sat outside on a bench, shaking and upset. What was happening to me?

As the summer progressed, I began to react to perfume, smoke, candles, dish liquid, and laundry detergent. I had an irregular heart beat, and great periods of weakness would come over me. This didn't frighten me as much as it pissed me off, however. I was pretty sick of having my life interrupted. I had become sensitive to many different chemicals and couldn't be around furniture stores, clothing stores, and some department stores. I couldn't visit one of my good friends because she had a gas stove. Hearing my son's band perform was out because the clubs where they played allowed smoking. Smoke had never bothered me before, but now I could not tolerate it. Diesel exhaust fumes and I became dire enemies; at times I would choke so badly that I needed to pull the car over and throw up. Prior to the printing press exposure at work in July 1994, none of these everyday chemical exposures bothered me at all. Now I began to need to restrict whom I rode with because new car upholstery and other things bothered me. As a result, I often went places alone, sat alone, and needed to leave early because I was feeling sick. As a "people person," I grieved about this involuntary social isolation and hated to accept it.

Our department management had at first embarked upon an earnest evaluation of the air quality problems in our building, but after a while they began to hedge, probably because they found out how much it would cost to fix the problems. They appointed lots of committees and spent a lot of money investigating the problems and possible solutions, but none of the significant solutions have been implemented. All they did was power vacuum the partitions in the rooms, clean the filters (which didn't introduce fresh air into our unit anyway), and wash the carpets.

Management continued to recommend that workers' compensation claims like mine be denied. One formerly productive employee who had become sick from this exposure retired early. Some of us who had no choice, like me, fought management. One member of management said to me one day in a testy sort of way, "You know, there are no bad people here." It wasn't about good or bad people, it was about ethics and accountability. It was about right and wrong. You don't ruin the health of over a half a dozen people from neglect and walk away from the situation as if it never happened. I couldn't walk away from it. I had to live with it. All any of us had done was go to work in good faith. We just wanted to be addressed with respect and not be manipulated. We wanted to be told the truth, allowed our dignity, and accommodated as our doctors asked. It would have been possible in some cases to accommodate this disability with isolation from problem substances through the use of tile floors, old furniture, and an open window.

I saw four doctors of my choice and one at the direction of workers' compensation. They all said my problems were work related. The doctor from workers' comp was just as direct as my specialist. Management and workers' compensation still hedged to avoid picking me up as a work-related injury. A lot of my co-workers observed the hassle I was getting, and it deterred them from trying to obtain accommodations they needed because they were afraid of losing their jobs or carrying a stigma.

My case worker told the doctors I was consulting that I would be moving to a new site where the simple accommodations I required would be honored. She was in fact moving me back to my old desk where not one accommodation had been addressed, but she led my doctors to believe it was a different site. I refused to go there, and once my doctors understood the situation, they wouldn't allow me to be placed there.

We had a mediation hearing to discuss my workers' comp claim, and my case worker wore strong perfume, even though part of the accommodation for me was that I would not be around perfume. It was February, and I was damned if I could wait three months for another hearing, so I opened all the windows in the meeting room so that I could stay there. She froze. I smiled.

I like to work, and I take pride in the job I do. I was honored as one of 4 "employees of the month" out of about 400 employees in our department statewide. After I began to work off-site because of my health problems, my boss said that the group of people I was supervising were doing so well that he was submitting us for consideration for the Commissioners' Team Award. Two of my employees told my boss that I was the best supervisor they had ever had. Shortly thereafter, I took a six-week medical leave at the recommendation of my primary physician. He thought that if I stayed away from the chemical exposures in my workplace for several weeks, I might desensitize enough to be able to tolerate the chemicals at work better.

When I returned to work, however, my boss wouldn't give me back my regular job duties. He said I was "at the end of the line" and there would be "no more accommodation." I filed a claim for discrimination against my boss under the American Disabilities Act and eventually won. He was demoted, supposedly voluntarily. He doesn't speak to me anymore, and if you think that bothers me, that's a big swing and a miss.

After that problem was resolved, I worked for a year and a half in another site that was wonderful (no chemical exposures that I couldn't tolerate). Unfortunately, the site became impossible for me when they moved in a huge photocopier, which meant I could no longer work in that site. I was out of work for a while but have now been placed in a site again. It's not working out too well, however, because I have a smokestack 100 feet from my window and I have to leave the window open to reduce my exposure to the low levels of chemicals offgassing from various things in my worksite. Management apologized that they had to put me so close to the smokestack, but their apologies don't go far toward improving my health problem.

Accountability is important, but our department doesn't seem to have recognized that. An example? Management still refers to this as the "alleged" air quality problem.

Marilyn's update: It is 1999 now and I have been working out of my home for a year. That has worked out well and enabled me to stay relatively healthy. Management considers this a temporary placement, however. Last April they remodeled what was supposed to be a safe room for me to work in, but that didn't work out for several reasons. I had to share a general rest room cleaned with strong chemical cleaners, I had to walk through newly carpeted rooms to reach my office, and I had to use a photocopier in another section. When I use a photocopier for any length of time, I have blurred vision and my breathing peak flow reading for asthma goes down about 30%. But the heat unit is the biggest offender in the "clean room." I am very reactive to it.

Since the failed work trial in the remodeled room, I feel I am in the "line of fire" from management. My supervisor actually stated he could not accept the idea that there was no cure for this disease. Since then my work has been scrutinized and "work problems" have been alluded to. This is a first for me, a far cry from the days when I was an "employee of the month" and had very good evaluations.

My level of health is directly related to what chemicals I come in contact with. A mild to moderate exposure brings moderate reactions that I can tolerate. A strong exposure incapacitates me. Working at home has enabled me to minimize my exposure to chemicals and keep my health from deteriorating further. I have spent much time defending this disability to my employer and fighting for my right to work at home. Because I feel I don't have another choice, I am filing with the Human Rights Commission this coming week based on a hostile work environment and disability discrimination. We'll see.

Tim

Gulf War Veteran

I was with one of the Army National Guard units that went to the Gulf War. I volunteered to go with this unit. I was a communications sergeant while I was there.

We arrived in Saudi in the port of Dhahran, spent two weeks in Dhahran doing different stuff, getting ready to deploy to the desert. Once we deployed to the desert, we went through little towns. Once the ground offensive started we were in Iraq and Kuwait.

Once when I was driving an officer at night, taking him to a meeting, we came upon a pile of dead sheep in the middle of the desert, not knowing what had caused it or how many there were. We didn't stop, we just turned the lights on and saw the sheep piled on both sides of this dirt road. The way that it caught our sight was that we saw the eyes of the desert rats reflected in our headlights, and we said, "What are all these things?" And then when we got close enough, we could see that they were eating the sheep. The night that we saw the dead sheep was the night of the first air offensive, and I was coming back from the medical facility, where I had taken somebody who had fallen, and en route back, we had gotten off the road that we usually took. Being out in the desert, all roads look the same, you know, you take a left here and a right there. That's when we came upon the dead sheep, and we didn't know if they were diseased or if there had been a chemical attack. Not being up on the front line at the time, it was like six hours that we had been gone, we didn't know what had transpired. That's one of the things that I remember vividly, seeing something that you don't normally see driving down through the desert.

On a nightly basis, we would spray our uniforms with pesticides. There was a chemical spray that they gave us to spray our uniforms. We had to hang them outside so that the excess spray would dissipate in the air, I guess. We weren't supposed to put them on immediately after spraying them. The sand fleas were a problem. We used to put flea collars around the legs of our cots or we would put flea powder on the floor around our cots to try to keep the sand fleas away from us while we were sleeping. We slept with nets over us to keep the flies off. The flies were ungodly. It wouldn't be nothing to go up to the mess hall and see a pile of flies where they had put some food out and let the flies land on it and then sprayed them to kill them. They would be like an inch thick, a pile of flies sitting on the ground outside the mess hall as you're going into the mess hall. It's the wonderful world of being in the army.

There were six men to a tent, and each tent had a kerosene heater in it. You would put it in the middle, and in the morning you would get up and light it and everyone would lie there until the tent got hot. But after a while, your eyes would start burning; there was an odor in the air constantly when you used that stuff. The oil wells burning was a huge problem. There were times when in the middle of the day it would look like midnight. One day at 11 or noon we got a wind storm, no rain, just wind, and everything went black. But it wasn't dust, it was just like a cloud had come over you and you couldn't see anything, like nighttime at noon. It was unreal.

You would look off in the distance and see the oil wells burning. You would see the smoke on the horizon, and it would just drift, whichever way the wind would blow it, you could see it drifting across the desert. You could smell it, you could constantly smell burning fuel oil and diesel fumes.

Most of the water you bathed in was unfit to put in your mouth. You couldn't brush your teeth with it. You had to use your bottled water to brush your teeth. The water was somewhat green, some of it had a tinge of diesel, a smell of diesel in it.

Well, my stomach problem, my gas reflux or whatever they want to describe it as, started while I was in Saudi. It was late, just about the time we were coming back it started, and it was diagnosed as gastric reflux. When I returned, we had a physical here, and they gave me a medication for it.

Other stuff that has come up is, I'm sensitive to the smell of gasoline. Diesel fuel, paint thinners bother me, other chemicals that I use. If I go to a party where there are a lot of women with perfume, it bothers me. Where I work I'm exposed to a lot of chemicals, and I seem to be supersensitive to them now. I can get a little smell of a paint and that will bother me. I'll fill up and my throat will get sore, whereas everybody else could just smell that, and it wouldn't bother them.

There was one instance when I first came back, I had been back maybe three weeks, and I was what they call a brazier. I would braze the pipe with silver solder and stuff. And I had developed bronchial pneumonia, and they couldn't figure out why. They cured me of it, but two or three weeks later I had pneumonia again. They did a test at the hospital, and it showed that the carbon monoxide level in my blood was three times what it was supposed to be. And from there, they determined that the job I was doing had become a risk for me and I had to change jobs, give up the job and switch into another section in the department.

Headaches, they're unreal. I take three medications for headaches. I take one before I go to bed that I can only take before I go to bed because of the strength of it, and I take one when I feel a headache coming on, and I take Tylenol. That's the three things I take for the headaches they say aren't related to my being part of Desert Storm. My doctor has tried three different medications for my headaches, which he says are cluster headaches.[1]

When I went off to Desert Storm, I left healthy, there was nothing seriously wrong with me. I had two kids I used to play with all the time, and now it's a lot different. It's work and rest and I'm not healthy now, it's cut and dried. There is something happened over there that they won't admit to.

Over half the people in the unit I was with have come down with something since they've come back that the authorities have said wasn't connected to their service in Saudi Arabia, but the normal person walking down the street in my hometown isn't going to come down with this. But what can you say, the government is either hiding something or they just don't want to let us know what's been done to us.

When I first came back, my prescription was $9 a month for my stomach problem. They wouldn't even say that was caused by Desert Storm. It was determined when I came back I had it, and I didn't have it when I left, but the government refuses to pay the $9 a month prescription on a monthly basis for me. They said it wasn't service connected. I've applied three or four times, I've reapplied and reapplied and they've refused it, said it wasn't service connected. I know how the guys with Agent Orange from Vietnam feel now, and I'm nothing compared to them. I'm $9 a month, I'm not thousands of dollars for treatment of the cancers and stuff that was caused by that.

There were maybe 12 or 13 of us in our unit who were volunteers, who stood up and said: "Hey, we'll go. You haven't got enough people. We'll go, we'll serve with you." And there was the government saying: "Thanks a lot. You went over there, we paid you while you were there, that's good enough. We don't need to do anything else for you. You're all set." And there were Vietnam veterans over there too with us. And they're sicker now than any of us. The government, they won't admit to anything. They're slowly letting stuff out, but it's not fast enough. Too many people are really getting sick and having to put themselves in debt to take care of their own medical problems.

---

[1] One medical encyclopedia states that the pain of cluster headaches is so strong that they are sometimes called "suicide headaches."

The illnesses that we've come down with, myself and the other veterans that went over with me in my unit, there's got to be a connection. The guy was healthy enough to pass the physical before he left. And to come back and hear about all the Gulf War illnesses and chemical sensitivity and everything that's come out of this, there's something there. The guy can't pass the physical to be in the military any more. They've got to prove it somehow. They've got an answer, but I guess they just don't want to give it to us.

The biggest thing is the sensitivity to stuff that you used to, you used to go out all the time and paint your house and use the thinners and stuff, and now you've got to avoid the use of the thinners. When you pump your own gas, like myself, I've got to turn away so I don't breathe the fumes. The chemical sensitivity is becoming just unreal, and you notice it now. Before, when you would pump gas, you used to stand there and smell the fumes, you know, great, this stuff don't bother me. And now you've got to try to hide and pump at the same time. And it's not just me, it's some of the vets that were with me over there. The same thing, and they're not getting paid for their medical bills either. And it's hard to stay serving your government when they don't want to tell you the truth.

Linda

Registered Nurse

I am a registered nurse and a social worker. When I went to work at a state-owned nursing home in Vermont in 1987, I enjoyed excellent health. After working there for about a year and a half, however, I noticed that shortly after entering the building, I would experience chest tightness and I would feel like I had a lump in my throat. I always felt like I was coming down with the flu. I had a headache every day I worked. I noticed that when I was away from the nursing home on the weekend, I would feel much better.

I started keeping a diary and recording my reactions: where I was, what triggered the reaction, and my symptoms. My diary indicated that I was reacting whenever cleaning was being done in the facility, which was almost constantly. I approached the administrator of the facility and told him that the cleaning products were making me ill. His response was not what I had hoped for. He was angry at me for complaining and said that the cleaning products didn't bother anyone else, so why should they bother me? Furthermore, he said the cleaning products had been bought in bulk and they would have to be used up before he would even consider changing them.

The reactions from my co-workers also shocked me. The housekeepers would not even look at me, much less speak to me. Staff would whisper to each other and laugh when I walked into the nurses' station. And these were professionals—nurses and social workers. One day I had a severe reaction that required a visit to my doctor, who recognized that I was having a reaction to cleaning chemicals. He wrote a letter saying that I should remain out of the building for two weeks to allow the residue of the cleaner in question to dissipate. He supported me in a workers' comp claim for the two weeks I would miss from work, and that claim was granted in December 1989. He also reported the facility to VOSHA (Vermont Occupational Safety and Health Administration) because he suspected that two different chemicals had been mixed together, creating a combination that caused me to react.

VOSHA's inspection found many things wrong, and recommendations were made to improve air quality through ventilation and the proper use and mixing of cleaning chemicals. I returned to work two weeks later with the naive expectation that I would now be able to work again with no physical problems. Boy, was I wrong! This was my first experience with how most Vermont state agencies work. Even though VOSHA had made recommendations, that did not mean that the facility was going to follow them. To my knowledge, VOSHA never followed up, despite the telephone calls from me, and when I left messages, no one called back. I continued to work even though I was having reactions every day while at work. At this point, I had no idea what was wrong with me. I had never heard of multiple chemical sensitivity. I thought I was going crazy, and this idea was reinforced by my co-workers, who thought I just wanted to complain.

Five months later, in May 1990, I had a severe reaction to a cleaner that I could identify. I took the label with me to the doctor, and he thought that I was reacting to the pine oil ingredient. He wrote a letter stating that I could not return to work for two weeks until all the pine oil had dissipated. The letter also stated that I had a limitation that prevented me from being around solvents containing pine oil. The facility received this statement from my doctor but never acted upon it. Once again, however, my claim for workers' comp was granted.

I continued to work and became very creative at avoiding areas of the building that were being cleaned. My job duties were changed so that I could spend more time in my office and less time on the floors of the facility, which were being cleaned almost constantly. I received permission to clean my own office, which gave me some control over my immediate environment. I continued to have some problems, but an eight-month maternity leave totally cleared my symptoms, and my previous state of health returned. My unexpected pregnancy at 43 years of age was a big surprise. Fortunately, my daughter was born healthy and has enjoyed good health throughout her eight years.

If I had known at that time about multiple chemical sensitivity and what I should have done to protect myself, I never would have returned to that facility after my maternity leave. But return I did, and the very day I returned to work, I started having reactions, even though I had had no reactions for the past eight months. To make matters worse, my reactions were now more severe than they had been previously. I developed reactive airway disease, where my neck swells, I feel like I can't swallow, and I lose my voice. I also become very short of breath, with chest pain and a sharp pain in the center of my upper back. In addition, I developed hives on my neck and chest. I could not believe that this was happening to me. In my ignorance of MCS, I wondered if I were having some sort of psychological incident. Once again, however, I could identify the cleaner to which I reacted, which happened to be a different one this time from that which had previously caused a problem. I went out again on a two-week workers' comp claim in December 1991. I returned to work on January 2, 1992, only to go out on workers' comp again on January 13. Whenever I had these two-week periods away from the facility, my symptoms would abate, and I would return to work hoping that everything would be fine, but it never was. I discovered that even though the facility had promised not to use the cleaner that had bothered me this time, staff members were using up what inventory they had on hand. This episode left me with a diagnosis of asthma, allergy, and hypersensitivity reaction.

One other woman had to leave the building the same day that I did because she reacted to the same cleaner. She was too intimidated to ask for workers' comp, however, and returned to work the next day. She has since been diagnosed with MCS. After my last workers' comp claim, the State of Vermont started to deny workers' comp claims based on allergies, which is how these reactions were being categorized. Another woman had a major incident in July 1999 after being exposed to the same troublesome cleaner. Then after being exposed to excessive perfume in October 1996, she was hospitalized for four days. She was having so much chest pain that she ended up in the telemetry unit so they could monitor her heart. A technician who came to give her an EKG was wearing so much perfume that it exacerbated all her symptoms. The patient's husband had to tell the technician that her perfume was making his wife sick, and he asked her to leave the room. None of the hospital staff really believed, however, that perfume exposures could be causing the patient's problems.

I continued to work and continued to have reactions on a regular basis. By this point, I could not tolerate the scents coming from perfumes, aftershave lotions, or cleaning products. My reactions were no longer limited to the facility where I worked but happened at the grocery store, in most public buildings, in friends' homes, and even in my own home. I had a teenage daughter who liked to use perfume and scented hair products, so my home life turned into a war zone like my work life. I think what really got me through this difficult period was having a young child to love and care for along with a supportive husband.

Meanwhile at work, two more of my co-workers developed MCS symptoms. We still had not heard of MCS, however, but were focusing on sick building syndrome. As we researched sick building syndrome, we found literature on MCS and discovered a physician in western Massachusetts who specializes in environmental medicine and has helped me tremendously.

By this time there were five of us from my facility who had been diagnosed with MCS. We were receiving support from the Vermont State Employees Association, the labor union of the Vermont state employees. If we had not had the union behind us, we know we would have been fired years ago. Several other employees would come to us to complain about problems they were experiencing with cleaners and the poor air quality. We would ask permission to use their names in our complaints, but they would refuse; they also declined to join us to try to improve working conditions.

We started asking other employees to not wear perfume, but to our dismay we could see that our comments only led to increased use of perfume. One of the women in our group of five was hospitalized for a few days in 1995 with a severe asthma attack after exposure to a heavy dose of perfume. When she went to the emergency room, her peak flow meter test was so low that the doctor told her that she had just escaped death and he was admitting her to the hospital. In the summer of 1995, I started having vision problems and was diagnosed with bilateral cataracts. I was only 47 and had no family history of early cataracts. Within the next few months, two more people in our group of five were diagnosed with cataracts. One was 45, the other 50. I had my first implant in September 1995 and my second in October 1998. I was very scared because I feared that if I reacted to the implant, my MCS would worsen and I might go blind. But because I only had minimal vision by this point, I decided to take a chance. Thank God, everything went fine. I have had no problems with the implants and enjoy great vision now. The ophthalmologist who saw all three of us stated that he was sure the chemicals being used at the nursing home had caused our cataracts, but he said that he would be unable to prove it. The State of Vermont refused to accept any responsibility for our cataracts.

By this point we were not the most popular people at the facility. Co-workers stopped speaking to us, and jokes were made at our expense. Then a new assistant administrator came on board who asked us if we were aware of internal e-mail messages that some of the women in the facility had been sending to one another about us on the company computers. We had not been aware of these messages, and when we requested copies, our request was denied. We then approached the union to get us copies, and the union succeeded in getting copies only by paying $130.

I find it hard to describe my emotions when I read the e-mail messages. I felt like I had been kicked in the stomach. After I recovered from the original shock, I started to cry. Reading how my co-workers conspired to wear heavy amounts of perfume, all the same kind on the same day, was horrifying. They even named the day according to the perfume they chose to wear that day; for example, one day was named Peach Petals day. They bragged about spraying the bathroom that we used with perfume and about spraying the top of the stairway that we used. They joked how all of us should dress up as "bubble people" for Halloween and they should dress up as cleaning products. One of the worst perfume offenders wrote on the e-mail, "Like I said before, shoot the bitches. I know where we can get some bullets." And this woman is a registered nurse.

We did not obtain the e-mail messages until September 1996. It happened that my mother had died on July 15, 1996. I remember working on July 14, 1996, so ill I didn't think I could survive because the perfume was so heavy that day. The nursing home that my mother resided in called me on July 14 to tell me that she might not survive the night. My husband begged me to go to the emergency room for myself because I was having such a hard time breathing. My lungs were so congested that he could hear my respirations across the room. When I went home before going to be with my mom, my little girl said, "Mommy, you stink like perfume." My co-workers had worn so much perfume that day that I had absorbed it in my hair and clothing. But I couldn't take time to go to the emergency room because I wanted to be with my mother as she was dying. As it turned out, the nurses at her nursing home worried more about me that night than about their patients. I cannot forget how I suffered that night, both from losing my mom and from the physical suffering that I later learned was the result of a malicious prank by my co-workers. When I read the e-mail and recognized the date of Peach Petals day as being the day I was called to the nursing home to be with my dying mother, I felt violated. My grief felt fresh all over again.

Throughout this period when I was having constant reactions, I repeatedly complained to VOSHA, but with no results. I applied several times to the State of Vermont's Reasonable Accommodation Committee for a reasonable accommodation under the ADA (American Disabilities Act), but my application was denied every time because MCS was not recognized as a disability. When I complained several times to Vermont's Department of Environmental Health, they did not even know what MCS was. Needless to say, they were no help. We complained to the State of Vermont's Risk Management Department, and they sent someone down to talk to us, but she said she couldn't understand how perfume could make us sick. We kept complaining to everyone we could think of, but we received absolutely no support from anyone in state government.

Finally, in August 1996, I filed a discrimination charge against the State of Vermont with the Human Rights Commission. I charged that I had been harassed, belittled, threatened, and made the brunt of jokes, as evidenced by the e-mail messages. I further charged that I was forced to work in a hostile environment, as evidenced by a co-worker wanting to "kill the bitches." In addition, I charged that my health problems were progressive and were exacerbated by the collusion of staff to purposely use scents that they knew adversely affected me.

In December 1996 the Human Rights Commission of the State of Vermont voted in favor of my claim, stating that I had been discriminated against on the basis of a disability—MCS. I was elated, but it turned out to be a hollow victory. Nothing changed where I worked. The same staff members continued to wear perfume. The housekeepers continued to clean with the same cleaning products. Air fresheners continued to hang in every bathroom and in many of the residents' rooms.

When I complained to the executive secretary of the Human Rights Commission, he informed me that the Human Rights Commission could not enforce its ruling. Attempted mediation achieved nothing, so finally the Commission issued us a right-to-sue letter, and we decided to pursue a civil lawsuit. Our civil suit is pending, and we have been given a court date of April 2000. We are suing the State of Vermont, the nursing home, and the individuals who were involved in the e-mail exchanges (who are being represented free of charge by Vermont's attorney general's office). Our group of five is now down to three. One woman dropped out because her husband did not want her to pursue the case; the publicity embarrassed him. The local newspaper heard of the story, and it was front-page news. Another woman in our group resigned her position at the nursing home but continues to be involved in the lawsuit. I greatly miss her daily presence at the nursing home because she was very aggressive in her pursuit of justice and was a great support to me. I have been left to be the leader, and I bear the brunt of the hostility at the nursing home. Several co-workers still don't speak to me. Invitations to luncheons and parties never seem to make it to my mailbox.

As I read what I have written, I know many people will think, "Why did she stay there? Why didn't she get out?" As I have previously stated, if I had known earlier what I know now or if I had been informed of MCS right from the beginning, I would have been out of there after my first reaction. I did not know, however, what was happening to me. I kept hoping things would improve. I was contacting what I thought were all the right agencies and people for help. I never anticipated that I would be ignored, lied to, and harassed. I never anticipated that my symptoms would worsen and one day not go away. I still work at this facility. I continue to fight every day for improvements, but it is a rare day that I don't have a reaction. I feel that I am now a prisoner at my job, held by financial considerations. I need to work. My MCS is always with me everywhere I go, so I doubt that I could work anywhere else without having reactions. If I did not have a union supporting me, I would probably be fired if I started complaining. There are not many job opportunities in the town I live in, and after all the publicity, I doubt anyone would hire me. I make an above average wage for this area and receive excellent benefits as a state employee. As long as I can work, I can add to my retirement fund. I am now 52 years old and have an eight-year-old child; I have responsibilities and I must go on.

Linda's update, September 1999: On August 1, 1999, the Vermont Veterans Home hired a new administrator. I was naturally very apprehensive. I thought that I would have to start all over again to try to educate him about MCS. I was concerned that he had already formed his opinion of me from information that he had received from the former administration and from the State of Vermont. After all, I am suing the place.

I was pleasantly surprised. He met with me and one other co-worker who has MCS. He told us that the past was over and that he could not change it. He said that he knew what MCS was and had dealt with it previously. He said that his philosophy was that all co-workers should treat each other like family members because we all spend so much time together. He assured us that we would not be subjected to the same behavior from co-workers as we had been in the past. We were told to report any use of perfume to him promptly and he would talk to the person involved. We have gone to him several times, and he has responded. He sent one woman home to wash off the perfume. When she returned still smelling of strong perfume, he sent her home to change her clothes. He has issued a perfume policy stating "The Vermont Veterans' Home discourages use of scented products and fragrances by all employees. . . . Management shall investigate complaints of the use of excessive scented products by employees when promptly presented with the complaint."

The new administrator has also hired someone with years of experience in the field of supervising housekeepers to find cleaning chemicals that will meet the needs of the facility but will not affect us adversely. We have an appointment to meet with this man to discuss cleaning chemicals and to inform him of products that are definitely a problem for us. This is a big step in the right direction. It's been hard hanging in there, so it's gratifying to see a major improvement in the situation at last.

Tony

Industrial Painter

I was raised down here in Cajun country, and down here we got the oil fields and offshore work, that kind of thing. Our main industry down here is oil. When I was 18, I went to work offshore. I started out as a galley hand, working in the kitchen. After a year or so, I started doing roustabout work on platforms, going around to different platforms. I did cleanup jobs, changed valves. Back then, I got a lot of exposures. I remember one time that a boat had accidently pumped diesel into our water tank on this rig. We had to drain the tank and get in there and clean it all out with chlorox and all, and it was pretty bad. It had that old diesel smell in there and all that, and that was one of the first things I did that was pretty bad. Another time I had to clean out the separator, which was full of fossil fuel, sand, and sea water.

For a couple of years, I did seismograph work on shore, and that was pretty clean. Then I came back down to the Gulf Coast area for one weekend, and I had an accident and broke my truck. I went to work with some friends of mine, a bunch of home boys from my hometown, as a sandblasting and painting helper. I was supposedly going to work just enough to get enough money to fix my vehicle and head back to Mississippi. I was pretty young and wild at that time, but I ended up working with the home boys and painting and sandblasting. There used to be a saying among painters that once you breathe enough MEK, methyl ethyl ketone, you won't want to leave, you won't want to do any other kind of work. That was pretty stupid.

Anyway, back then, there were very few regulations or anything about safety or using respirators. We used to paint with T-shirts over our heads. I worked in the painting industry for two years before I even knew what a respirator was. We used these desert hoods, little canvas hoods to sandblast. We were breathing a lot of dust and sand, and everybody was scared of getting silicosis of the lungs. We hadn't really focused that much on how dangerous the paints were. We knew that when you read the back end of the can, it said it'll cause this and cause that. We used to joke and say, "Why don't they just say it'll kill you?" Anyway, we never paid too much attention to the back of the label. They weren't very strict about safety back then.

So anyway I stayed working with these home boys and got caught up in sandblasting and painting and started working offshore; mainly that's where I did most of my work was offshore. One time we went out to an eight-legged platform with a crew of six men, and we sandblasted and painted that whole thing from one end to the other in 103 days. We used to boast about ourselves, about what good painters we were.

Anyway, I was always breathing MEK, which was a cleaning solvent we used every day. We used it morning, noon, and night to clean our equipment. It was just part of our life. We washed our clothes together, and sometimes we lived in little bitty bunkhouses that they moved from rig to rig, like a small trailer with absolutely no windows. You get a gang of sandblasters and painters living in there with painted-up clothes and all of that, and it's just like you're living in constant fumes.

I went underneath cellar decks on platforms and stuff. I walked on 3/8" cables, holding on with one hand, and nothing below me but iron and water. Anyway, there were very few safety concerns. It was very seldom that anyone wore safety belts. We used to have a joke that if you fell, if you were lucky, you would hit one of those pipes and it would slow you down before you hit that salt water, because that salt water was hard. I'm not no John Wayne or nothing, but there's not too damn much that I'm afraid of.

I worked on boats where we had all our equipment on the boats and we would back the boats up to the platforms. I've been around diesel motors and stuff—compressors that were so loud you had to scream at each other to hear each other, which was really bad for our ears and all, and we would sleep on the boats. Every one of those boats at one time or another had got their water tanks and their fuel tanks mixed up because on almost every boat in the Gulf of Mexico, when you took a shower, you could smell the diesel. That's a harsh statement, but most of the work boats I worked on were like that. It may be different today. This was some years back.

Finally we hit the oil crunch in the early 1980s, and I started taking work in different plants. I landed a job which was to be my last job as being a part of the workforce. This was around Christmas of 1988. I landed a job with someone who wanted somebody who didn't want to work offshore. Everybody was screaming to work offshore, but I had had my fill of offshore work. I wanted to work on land. It was a matter of being in the right place at the right time. I ended up with this job that was a really good job. I worked there as a sandblaster and painter at first. Then they put me in charge of buying paint and stuff. I changed the paint system around. It was a good job, and slowly, little by little, I started doing a little bit of truck driving, here and there. Then the old guy that was the yard hand, head gofer, had a heart attack and had to quit. They kind of in a roundabout way gave me that job. I started getting away from the sandblasting and painting little by little. I would paint a little, but I was driving trucks more than painting. They changed my job title, moved me up, everything was hunky-dory.

So everything was going pretty good when this guy came over to the boatyard and wanted his boat painted. It was a small bass boat that was homemade. He had one gallon of automobile paint that he wanted to use. I had never touched a drop of automobile paint in my life, and I'd been sandblasting and painting for twelve years.

Now prior to painting this boat, when I was getting away from sandblasting and painting, gradually getting away, I was beginning to notice that every time I did spray paint, and I sprayed in particular urethane paint containing isocyanates, I was having problems. I had to have a respirator just to mix the paint, which is a common safety practice now. Usually after we got it mixed, we would try to spray a little bit with our gun to make sure that it was working correctly and all that, and then we would go cover ourselves up real good and put on our respirator. But it got to the point that I had to have my respirator on just to try out my spray gun. These urethane paints were beginning to bother me, and when I went into the shed where all our paints were stored, I was having problems in there. When I'd get away from it, I would start to feel pretty good. Then around the back of our paint shed we had MEK stored, and it got to where I couldn't tolerate the MEK. I was having this sort of like a pinching sensation in my chest anytime I got around MEK or urethane paints.

We used to have safety meetings, and they would say things like: "Don't slip, don't trip. If you go up a ladder, hold onto the hand rail." No one ever talked about what kind of symptoms you would be having if you were beginning to develop a loss of tolerance for all these chemicals.

Anyway, that guy with the boat kept coming around, coming around, saying, "When are you going to paint my boat?" We were waiting for the weather to get right, and there was a lot of stress built up.

So the day rolls around to paint the boat. I had to use the welders' shop to paint his boat because I had to have it suspended in midair. That Sunday afternoon I started painting the boat, but it went on into the night, and I was painting some other things outside and going inside to paint the boat. I'd go in and out, in and out of the welding shop that didn't have any kind of ventilation because there was sand everywhere and the wind would blow the sand, and the man didn't want sand to get on his pretty boat. He wanted it done to perfection. So going in and out, in and out, putting my respirator on and off, painting outside and painting inside, I think my body just reached a level where that was it. I crossed some sort of a line.

I felt a little strange that night when it was all over, a little wobbly. When I got in my car to drive home, I felt kind of bad. I went and picked up my children at the babysitter's and went home and slept. When I woke up the next morning, it felt like I had kind of a chest cold. On Tuesday it felt like I was kind of coming down with the flu, and I noticed that when I would exert myself, I started to feel real bad, but I kept trying to work. I was a good, loyal employee, went to work every day, I was never late. I never minded putting in overtime. In fact, up to the point I painted that boat, I had worked 21 consecutive days nonstop, every day, so that may have been a factor in my body being overburdened with these chemicals and all.

On Wednesday, I had to go to Texas and deliver some welding equipment in a great big truck that ran on diesel, and it felt like I had a block of ice on my chest. When I was coming back, I was having a hard time driving; it still felt like I had ice on my chest. Then on Thursday morning I drove halfway to work and said, "Man, there's something wrong with me." By this time, my ankles began to feel like I was walking around on sprung ankles, and the bottoms of my feet were like on fire. I just knew something was terribly wrong, so I turned around and went to a doctor. He was an old country doctor, he was a good old boy. He checked me out, and he said, "Man, you've got pneumonia in one lung." He gave me a shot and some medication and sent me home.

Friday morning I felt like I was dying, and I went back to the doctor and he checked me out. He said, "You've got bilateral pneumonia. I'm going to have to put you in the hospital to clear this up." When I went to the hospital, a woman who worked there whose husband was a friend of mine at work saw me going in the hospital. She called up the people that I work for and told them, "Man, you'd better come and see about this guy. He looks like he's fixing to die." And I did look like walking death going into that hospital.

But by 5 o'clock that afternoon, I looked like the perfect picture of health. I didn't look like there was a thing wrong with me. The doctor came in when he made his rounds, and he said, "You sure bounce back fast." I said I was feeling better. I hadn't really put my finger on it that my house was so chemically saturated because of the kind of work that I had done, dragging that stuff home. I guess it helped me to get in the semiclean hospital environment that was a whole lot cleaner than what was at my job or my home. The doctor said he couldn't figure out what was wrong with me, so he kept me in the hospital the rest of the weekend to run a few more tests. He let me out late Sunday afternoon. I went home, but by about 8 or 9 o'clock that night I was going downhill, I was in a bind all over again.

Monday morning I went back to the doctor's office, and he took one look at me, and he scratched his head and said, "I don't know what's the matter with you, but you look like somebody had beat the hell out of you." I didn't know anything about reactive airway disease yet. I hadn't made the connection that when I had painted that boat inside, my body just met with its tolerance level. I hadn't really put all the stuff together. The doctor said, "Go home and sit and rest. I don't understand it. I've given you so many antibiotics, you shouldn't have a germ in your body." So he made me an appointment to go see a pulmonary function specialist.

When I went to see the pulmonary function specialist, he started backing up in the sequence of events, when all this started, and that's when I put my finger on painting that boat. Painting indoors is a lot different than painting outdoors. Anyway, he ordered a SPECT test for me, and he made me an appointment with a neurologist.

In the meantime, I was taking pain medication. I couldn't walk from the sofa to the bathroom. I was really in a bind because I'm a single parent. I had to sit in a chair to wash the dishes and stuff. I just literally couldn't hardly do anything.

I was bound and determined, however, to get well and go back to work. People kept calling me to go back to work. They would say, "When are you going to get better? The work's piling up, we need you down here." All that kind of stuff. Finally, I was getting hungry. It had been about three weeks, I was running out of money and all that. So I went to the doctor and I said, "Man, are you going to let me go back to work?" He was totally against it and said, "I'll let you go back to work, but absolutely no painting and extremely light duties. Hang around the office and make calls."

So I went back to work on those terms. But by the second day I was there, they had me loading a truck to go to Texas, the same diesel truck. I found that when I would exert myself, my body would pound, much like if you smashed your finger with a hammer. Finally, I told them, I said, "Man, you're going to have to come and help me with this truck." So they came out there and helped me load. I ended up driving it again. When I came back home that night on the interstate, man, I thought I was going to die. The next day I went into work and said, "I just can't do this, man." About that time the pulmonary function specialist called me with the results of that SPECT scan. He said, "You've got slight brain damage, but with the right medication and physical therapy, you'll be all right in about six months." I left work early and went over to have this guy tell me this face to face.

So I took this all to my boss and told him that I was sick like this and that it was going to take me a while to get well, but I still wanted my job, I still wanted to work. I loved my job, I wasn't lazy. The last thing I was,

was lazy. He said, "Well, we'll fill out an accident report," and he started turning it in to the workers' comp people and all that, and everybody started running like ants because it was a big thing if anybody was ever to get hurt right there on that yard. It was one thing if somebody got hurt offshore where the big bosses didn't have any control over it, but to get hurt right there on the yard with all these big bosses around, somebody was going to take a fall. Anyway, what happened was, they didn't want to acknowledge I was hurt. My boss crossed his arms, and he stood in the corner, he looked at the road, he looked at me. He said, "If you can prove the paint did this to you, I guess we'll have to take care of you." And I looked at him and he looked at the road, and he looked at me and I looked at him, and he looked at the road and that was all.

From that moment on, nobody down there even wanted to talk to me. I kept trying to talk with the main office to try and get my workers' comp, and the workers' comp man called me on the phone. He claimed that I didn't want to get along with him, and he said they had reviewed my case and they didn't think I was hurt. Anyway, they denied me my workers' comp. So I went to the bossman who was the owner and a Christian kind of a guy. I talked to him, and he called the workers' comp guy back and told the workers' comp man, "Look, Anthony is an exceptional employee, and he should be treated as such." But the workers' comp people still treated me like they treat everybody. I didn't get no workers' comp. I had my private insurance that I was paying for out of my payroll check. I kept going to doctors with my private insurance, trying to find out what was wrong with me. I went to this neurologist, and he said, "You look OK, you can go back to work." I said, "I'm telling you I can't walk around my house without having some place to sit down." Anyway, it turned out he was working for the insurance people or whatever. I hate for things to get ugly, but a situation like this gets ugly.

My private insurance wouldn't pay for anything because they thought it was a work-related injury, and the workers' comp wouldn't pay because they thought that I wasn't hurt, so I kept doctor-hopping around. I ended up in a charity university hospital and spent 17 hours waiting to see a doctor. Here I supposedly had private insurance and should have had workers' comp and neither one of them wanted to pay. There was nothing I could do. I threw my hands up in the air.

Then I stumbled across a doctor who said, "I'm not the one who can help you, but I've got a pretty good idea of who can." He sent me to this other doctor who was a specialist in occupational toxicology and occupational illness. I went to see him with SPECT scan in hand. The guy knew right away what was the matter with me, and he began to treat me. By this time I had gotten a lawyer. He was forcing my insurance company to pay to take care of me; otherwise, they might end up with a wrongful death case.

I ended up having endoscopic sinus surgery, and the biopsies showed I had reactive airways disease. All the pollution down here just makes that worse. There's no place to breath clean air around here, there's no more fresh air, especially in Louisiana and in the house where I was living.

I was battling workers' comp all this time on the legal front. I went to a total of 17 depositions, and the very first deposition that I went to was in my boss's office down there almost on the Gulf of Mexico. Every one of my fellow employees came in there and verified that I was an honest, hard worker, had a nice personality, was always on time, and they loved working with me. Not a one of them said anything bad. The workers' comp lawyer said at that point that they realized I was hurt in the course and scope of my job and they were going to take care of me. They were going to start giving me 2/3 of my salary.

But what the workers' comp did was they didn't like what that lawyer told them, so they changed lawyers and denied the whole thing, and I started in for the fight of my life. After that I went to an additional 16 depositions. I spent three days in court. Finally on December 3, 1992, a year and nine months after I got hurt, the judge decided that they were responsible for my injury. The way that happened was that to prove to this bunch of idiots that I was hurt, the doctor had a nerve taken out of my ankle, a nerve that you supposedly don't really need, and had it analyzed, had biopsies done on it to prove I had nerve damage throughout my body and especially in my extremities, my hands and feet. This was sent to an independent medical examiner who had a name a foot long. This guy was president of some big organization, a specialist in toxicology and all, and this guy had decided, yes, the man was injured with the painting chemicals and stuff. When he sent his report to the workers' comp, he put in ten pages of references to old painters and what happens to old painters and how their bodies just get all tore up.

The workers' comp people were still thinking about appealing after all of that. By this time they had pretty much beat me to the ground. My car had four bald tires, there was one brake on one wheel. My children were running around with holes in their shoes and pants. I was on welfare, food stamps. My landlord was letting me live on credit, my babysitter was babysitting on credit.

My lawyer said the judge had charged the workers' comp people a $3,000 arbitrary and capricious behavior penalty for what they had done to me. But $3,000 was nothing for what those people put me though when them people knew after the first deposition that I was hurt. They knew they had so much time to keep their money in the bank.

Finally, on February 14, 1993, I got the first penny out of them guys, almost two years after I was hurt. So I walked into a drugstore and said to the druggist, "I'm not on private insurance anymore, the battle's been won." Or so I thought—the sigh of relief was really shortlived. I told the druggist that workers' comp was going to be responsible for paying for my medications, and I handed him the prescription. He called the workers' comp guy, who said there was not sufficient documentation, we need this, we need that, and all hell broke lose. For the next nine months, the rest of that year, I was having to pay for my medication out of my own pocket, I wasn't getting reimbursed. What them people did to me in the next nine months was almost as bad as what the chemicals had done to me, and that's when I coined the phrase "battered insurance syndrome." It got so bad dealing with those people that it was a trauma in itself just to look in their direction.

My poor doctor had been giving them people documentation on top of documentation. They had some nurse working in New Orleans debating whether I needed this or whether I needed that. So slowly I started weaning myself off of the medications and going toward the health food store.

When I was first hurt, I started losing sleep right away. When I laid in bed, my legs were just twisting and turning. I would sleep only two or three hours a night, and when my insomnia was at its worst, I would sleep only every other night. I stayed like that for four years. My doctor was giving me knock-out pills, and he said he couldn't understand why I couldn't sleep because the pills should have knocked me out. It wasn't until I moved and began to clean up my environment, get myself a safe place to live, that I began being able to sleep.

After I first became sick, I was bound and determined to keep living down there by the Gulf, but that didn't work out. My neighbor would see mosquitoes in her grass when she was mowing it, and she would call the town officials. They would come out with something that looked like a leaf blower and blow pesticide all around. I couldn't take a walk at all. I decided to move further north away from the coast, so I moved into this trailer that was twelve years old. It has sheetrock walls. The floor was particleboard, so I sealed it. It took a couple of weeks before I could sleep in the bedroom.

By this time I had thrown away almost everything that I owned. When you come to my house, it looks like somebody is just moving in or just moving out. We live with very little, but we get by. My children learned to adjust. It was hard, but they've done it. They tell me that in some ways they're better for it. It's taught them some valuable lessons and all in kind of an underlying way. Anyway, I went through a lot of trouble with neighbors and all. They would see me hanging things out on the clothesline for two or three weeks. People just don't understand how long you have to air some things out.

It's almost amazing that a guy like me who's been around chemicals, been exposed to raw fossil fuel, diesel, and oils and all that stuff—I smelt it all—and today if I walk out my front door, and my neighbor has got her dryer going with those fabric softeners or harsh detergents, it'll chase me right back into my house.

If I go roaming around and trying to act like a regular person, I'll come back home with what I call a chemical hangover. I'll have trouble breathing, my feet will hurt, my whole body hurts. And in the morning, man, it takes everything I got just to get out of bed. That's when I kind of came up with these analogies to help me understand my illness better, like fumeaholic, chemicalism, where I started integrating these things because to me this is a way of putting it on a social level. You don't have to open up a big two-foot book to try to explain why this stuff bothers me, you know. This is a tolerance thing, it's all about what you can tolerate. You know, I heard some guy on television talking about how this MCS is just an odor aversion. Man, that just takes the cake.

Kelly

Registered Nurse

The year 1996 was a horrendous year in my family's life. I am grateful I'm alive to tell our story and hopefully prevent others from going through what we have been through.

In 1993 my husband Russ and I and our daughters Keri and Kayla moved into an older brick home right across from a dairy farm that my husband had recently started working at. I had just graduated from nursing school and was excited to begin my new career. By 1994, however, I started noticing some vague symptoms—mainly fatigue, headaches that started immediately after I woke up, numbness in my feet, and urinary frequency. I went to see a doctor to have some blood tests run, and everything came back normal. I just blamed everything on stress from my new job as a staff nurse. After a year, however, I felt that I could no longer mentally or physically handle my job as a staff nurse. When I had a problem remembering words, I knew I had to get out of that environment.

My co-workers and friends kept telling me that after you turn 30 your body falls apart. I hated to accept that theory because I had always been an athletic person. But I had no energy left, so I found a less physically demanding job as a home health nurse in 1995.

By this time, other family members were also having problems. The girls and I were having vision problems, and they were fitted with glasses for the first time. Kayla, who was 10, developed a hacking bark of a cough that would never quit. She also started gaining weight. It was a daily struggle to get her to go outside and play because she wanted to lie around all the time. Keri, who was 13, began having headaches, and she too began to put on weight and didn't have a lot of energy. She also developed chronic diarrhea.

My husband Russ began to change also. Although he had been a workaholic all his life, he began sleeping every chance he could get. He became irritable and experienced frequent mood swings. When he was exposed to diesel fuel or exhaust, he would get rude and mean and he would scream at us. Later he would apologize. And he also started getting headaches, although he had rarely had headaches before moving into this home. In March 1996 we all flew to Iowa to help my grandparents celebrate their 70th wedding anniversary. We were embarrassed that my 90-year-old grandparents danced to every polka while we sat and rested because we would become so short of breath.

In May 1996, Kayla experienced a period of incontinence. Thinking that was unusual for an 11 year old, I took her vital signs and listened to her heart rate. I became frightened because it was irregular, so I took her to a pediatrician. He just told me that the irregular heartbeat was normal for children and she would grow out of it.

In the meantime, my own symptoms had gotten worse. I now had difficulty walking up steps or inclines as I would become very short of breath. I started noticing problems with different odors like those from air fresheners and perfumes.

Russ had been seeing a chiropractor now for the past few months because he had joint pain, stiffness, a sore back, and a stiff neck. By the end of May, he experienced severe headaches, had problems thinking straight, and developed a trunk rash. I was finally able to talk him into seeing a doctor. He went in and had his blood drawn, only to return again the next day because his arm started going numb and he had some leg tremors. Fearing he was having a stroke, I took him to the emergency room, where they admitted him to the hospital. I had to stay with him to answer the doctor's questions because Russ couldn't remember anything at this point. He spent four days at the hospital, but test after test came back normal. He eventually got to the point where he could walk again without falling and was discharged from the hospital. Russ came home feeling better even though he had not been treated with anything besides a medication for his headaches. But within two and a half weeks he got worse and landed back in the hospital. He was having a severe headache, became very short of breath, complained of ringing in his ears, and was unsteady on his feet. (He would fall backwards.) He was also experiencing leg and arm tremors, skin discoloration, numbness of his arms and hands, and chest heaviness. The doctor said, "We will leave no stone unturned," and I thought that this time they were going to find out what was going on.

Our mechanic, who is a personal friend, heard Russ was in the hospital and stopped by to visit. When I explained what symptoms Russ had, he said he wondered if Russ had had a mini-stroke because the symptoms were identical to his own and he had been diagnosed with a mini-stroke earlier that year. I told him that Russ was in one of the best hospitals in the country and surely they would diagnose his problem soon. Later that week, however, the physicians told us that they were unable to make a diagnosis and that Russ would be discharged the next day, with a doctor following up closely on him as more signs and symptoms developed.

When I went home that night, I stopped to pick up our girls at the home of one of my friends. While we were all sitting in the living room talking, Keri had what appeared to be a petit mal seizure. All of a sudden she said,

"Roger, don't go," which made no sense at all because there was no Roger in the room. Then she got a blank stare on her face for a brief moment. When she came to, she asked what had happened. When we told her, she said, "This has been happening a lot to me at school lately and it is so embarrassing."

When the girls and I walked into our home that night, I noticed a fuel oil smell, the same smell I had been smelling on and off for the last few years. We had never taken the smell seriously before because we heated with fuel oil and figured the odor was normal. But then I started wondering why we would be smelling fuel oil in the middle of the summer, so I decided to have the smell checked out the next day.

The next morning our landlord and I went to investigate under our home. When I opened the crawl space door, I was almost knocked over by the fumes that came rolling out. Shining the light, we could see fuel oil standing in a large pool at least eight feet in diameter directly under our furnace. Apparently, the filter had been slowly leaking all these years. I ran in to call our fuel oil company to see if this could be what was causing all our problems, but the man I talked to just laughed at me and said, "Ma'am, there's nothing wrong with breathing fuel oil."

Next I called the poison control center. The man I spoke to there was a hero in my mind. He explained hydrocarbon poisoning and said a lot of doctors are unaware of it. He graciously agreed to speak with my husband's doctor but forewarned me that he didn't think the doctor would call him for advice because they rarely did.

Next I called Russ's doctor and said, "I've solved the problem." The doctor's response? "Ma'am, we have never heard of that happening before." I also explained that Keri had just had what resembled a petit mal seizure. I got the impression that the doctor just felt I was in denial over Russ's undiagnosed and potentially life-threatening neurological problem. They discharged him later that evening with a prescription for an antidepressant and a diagnosis of "general fatigue." (Nine days total in the hospital to come up with that diagnosis?) Unfortunately, I was unable to pick Russ up that night because I myself had become severely sick from inhaling all those oil fumes that morning.

That summer we ended up excavating from under our house a section of earth that was soaked in heating fuel oil to a depth of at least three feet. The process took almost two months. When we tried to move back in, we were still sick, but by this point, even when we were out of the home we were ill. The doctors were of no help; I don't know whether they didn't believe us or just didn't know what to do for us.

We finally spoke to a physician with the Department of Environment, Health, and Natural Resources about the heightened sensitivity we had all developed to everyday chemicals like perfumes, air fresheners, diesel exhaust, fresh tar on roads, and household cleaning chemicals. We told the doctor that we would get reactions when we were exposed to these substances. He just replied that a lot of doctors don't believe in that and call it "power of suggestion" thinking. I didn't know what he was talking about at the time.

Then in January 1997, we watched a *20/20* episode with John Stossel that dealt with MCS. I had never heard of MCS before, and I nearly jumped out of my chair as I said to my husband, "That is what we have!" It was actually comforting to know there were others out there like us, even though John Stossel portrayed them in an unfavorable way.

During this period, I took our car in to be worked on by our mechanic friend at the garage in town. When I went to pick it up, I could barely breathe in their shop. I told them how unhealthy it was for them and updated the clerk and mechanic on my husband and our house situation with the oil spill. (I was hoping they would see a connection.) They told me that my car wasn't finished yet because almost everyone in the garage had called in sick. They then mentioned that four guys in the shop had been diagnosed with mini-strokes and that one had actually died two years earlier. I offered to talk to their boss and bring in some information I had received on the Internet as well as my copy of our material safety data sheet on fuel oil. They both quickly said, "Oh, no, don't talk to our boss; we could lose our jobs." At least I took some information in and gave it to the mechanic. (Update on the clerk two years later? He just underwent surgery for a brain tumor this past summer.)

During the summer of 1996, when we were having the oil-soaked soil removed from under our house, my 27-year-old brother died. Early that summer he had said to me, "I think I have what you guys have." He had served in the Gulf War, and he was very ill with Gulf War syndrome, but he received no help from the government. His death may have been the result of taking two morphine pills to kill the constant pain he was experiencing. (He obtained the pills from his terminally ill brother.) The autopsy showed no signs that he had overdosed on any other drug. After my brother's death, we were finally able to locate his lost records, and we read an entry in which a doctor had written, "Patient appears to be malingering his illness for secondary gain." I was amazed when I later learned that many other veterans had seen that same phrase used in their records. So now I have two missions in life. One is to help educate people about the potential adverse effects from the use of all the chemicals in our lives, and the second is to help support sick veterans and let them know their illnesses are "not just in their head," as some are being led to believe.

I am happy to report that our family is basically a healthy family again, and we no longer have that heightened sensitivity to chemicals that lasted over two years. We have drastically changed the way we live, however. No more harsh cleaning chemicals, no hanging around in department stores or hardware stores. If we smell a perfume at church or a meeting, we get up and change seats. It has been a long time since I have experienced numbness around my nose or lips when I am exposed to some chemical. The "brain fog" is still present at times when I have been exposed to something, but at least I don't have the slurred speech symptom that used to bother me. I'm still concerned that with all that exposure to heating oil, one of us may end up some day with cancer or a tumor, but at least we no longer live in the diesel tank we used to call home.

Carl

Gulf War Veteran

Before I went to the Gulf War, I had a full-time job as a carpenter. I was also the pastor of a Baptist church and the chaplain of my Army National Guard unit. My health was good before I went to the Persian Gulf.

My unit arrived in Saudi Arabia in early December and soon moved northwest to our desert camp about 15 miles from the border of Iraq. When the air war started, chemical alarms went off many times. This was the point at which we were given pyridostigmine bromide pills because they were supposed to be an anti-nerve gas medication. Within a few days, I wasn't feeling well; I was fatigued, had headaches, and muscle and joint aches. At this time, I was so busy doing lots of different things that I thought my health problems came from being on the go for what seemed like 24 hours a day. Within a few weeks, however, I started noticing that diesel exhaust and the fumes from diesel fuel were making me slightly nauseated, even though I had been around diesel before the war without any problem.

When the ground war started, I went forward with my company to a position only five miles from the Iraqi border. We had one scud missile explode about seven or eight miles from us. After Saddam Hussein gave orders to set all the oil wells on fire, the air was so full of smoke that daytime was like nighttime. We were close enough to the fires that we could see the flames and hear the roar of the fires. The smoke started bothering me a lot.

The insect spray we were using was also bothering me. My skin would get red when I used it. By this point, I was often getting short of breath—I didn't know whether that was from the fires or the insect repellant. I had bronchitis several times while I was in the Persian Gulf. Before we left Saudi Arabia, we were given a physical. My blood pressure was really high, 180/104. Before I went to the Gulf War, it had been 120/70.

In June 1991, I was released from active duty. A week later I ended up going to the Veterans Administration hospital in my state. I was having a hard time breathing, and I had bronchitis again. I told the doctors there about the various other symptoms I was also having. By this point I was getting sick from household cleaning products, and going into a store would make me sick. Before the Gulf War, I had only had a couple of migraines, but during the war I had started having them very frequently, and they continued when I returned to the States.

Since my health problems continued for a year or two, the local VA hospital finally sent me to the VA hospital in Washington, D.C. There they diagnosed me with chronic fatigue syndrome, asthma, sleep apnea, and brain damage (the brain damage was shown on an MRI). My hands and arms would also sometimes go numb, and they told me I had carpal tunnel syndrome. I'd never had that before the war, however. While I was at the VA hospital in Washington, I met other Gulf War vets who had symptoms similar to mine and were also sensitive to chemicals at this point. Through them I learned about a doctor at a VA hospital in Northampton, Massachusetts, who was very interested in multiple chemical sensitivity. About six months after I left Washington, I entered the Northampton hospital, where they kept me for three months and I was diagnosed with MCS. They sent me to a VA hospital in Connecticut to have a SPECT scan. It showed damage to my right frontal lobe and right temple, as well as damage to the right side of my thalamus. The results of the brain scan probably explain the memory problems I've had since the war.

I have had to make many adjustments in my life because of the chemical sensitivity I brought back from the Gulf War. I had to give up my job as pastor at my church because I was having a hard time remembering scripture passages that I used to know by heart.

Because the construction trade was slow at the time I returned from the war, I got a job in law enforcement because I had previous experience in the field. I worked for the sheriff's department for a while and also for a city police department. I wasn't able to continue that line of work, however, because I had become very forgetful since my return from the war and would get confused about where I was. There would be times when I'd be out driving on roads that I'd traveled many times before, and sometimes I couldn't remember where I was. I would just have to keep going, and finally I would realize where I was. When I was trying to work in law enforcement, I would get a call to go to a certain location where something was happening, and I couldn't remember where that street was located. So I just had to give up that work.

Before I went to the Gulf War, I used to love to go snowmobiling, but I can't go anymore because the diesel exhaust fumes give me a headache and make me feel nauseated. My sensitivity to chemicals makes life pretty difficult. Perfume bothers me, and so do newspapers and magazines. I get nausea and headaches from the ink, and my nose gets stuffed up. I broke out in rashes while I was in Desert Storm, and I still get rashes today. I have a lot of problem with asthma now, and I didn't have asthma before the war.

MCS also affects your home life a lot. It's hard on my wife to have me be so sensitive to chemicals now. When I left for the war, I was in good health. Now my health is bad, and I never feel like myself, I never feel good. There are many places I can't go because people are wearing perfume. Sometimes I have to wear a mask to go in a store, and people often stare at me when I wear it.

At first I was having a hard time getting compensation from the government when I filed claims on the basis of different medical conditions I had been diagnosed with. They kept turning me down. Then finally they used some other diagnoses, and I was able to get some disability payments. I'm now getting 100% compensation because I can't work, but I had to keep pushing and pushing to get that money. I just wouldn't give up, and the Disabled American Veterans Association helped me out a lot. It took me seven years to get that full disability status from the Veterans Administration. In the meantime I had applied for Social Security disability in 1996, and I got that in seven months.

I've noticed that the younger servicemen who are Gulf War vets are reluctant to let people know that they're not feeling well. Many of them are still on active duty, and they're afraid that if they let anyone know that they're sick, they will be discharged.

The VA hospitals are still just treating symptoms, but they haven't come up with any cure. I think it's a terrible shame the way the Gulf War vets have been treated, but it's no surprise to me because the Vietnam vets who were exposed to Agent Orange were treated the same way.

Lizbeth

Computer Systems Engineer

This is so very tough, talking about the nightmare that started almost six years ago. Every time I start to think about it, talk about it with someone, I get so angry that I could put my fist through the ceiling.

I have thought about this anger a lot lately. Mostly because it has stopped me from telling my MCS story. And I guess, depending on how sick I am on the day you talk to me about it, the degree of fury you have to deal with will vary. I know from hearing their stories that some people actually develop MCS just going through normal, everyday life. But in my case, I developed chemical sensitivity in one workplace and it was exacerbated in other workplaces. My employers bear a heavy responsibility for what has happened to me, but that's not a responsibility they seem willing to accept.

So here goes the fist through the ceiling . . .

My problems started on November 3, 1993. I will never forget that Monday, the last normal Monday of my life. The large company where I was employed as a computer specialist moved my work unit to a renovated space in a building called the Data Center. The Data Center didn't look too bad on the outside; it was a one-story brick building. But after you opened the front door and went through the small lobby, you were hit by the darkness of the place. I remember walking down the long corridor through a couple of sets of security doors. Then, on the left, I passed through another set of heavy doors.

As I walked through the last set of heavy doors that first day, I was struck by the sheer ugliness of the place. They had laid down dark, new industrial carpeting. There was a narrow walkway between two long rows of cubicles with new particleboard dividers. To our right were all the mainframe computers. The walls had just been painted a dark burgundy. And the air was heavy, really heavy feeling. There wasn't a single window.

By the end of the first week, on top of the place really smelling strange, two of my co-workers and I had odd feelings in our throats. For want of a better description, it was like having fur balls in your throat. My throat kind of scratched and hurt; it felt like there was something there, even though there wasn't.

Within ten days after I entered the new workplace, I started hacking late one afternoon. Not really coughing, but what I later learned is called wheezing. My workmates weren't doing too well either. A lot of them were having a variety of classic allergic reactions like sneezing, watery eyes, and itching. I had also suddenly developed a prominent rash on my elbows and knees that got so bad it resembled eczema.

By the time I got home about 5:30 P.M., I was hacking so badly I had to hold onto the kitchen counter to keep from falling on the floor. I had so much trouble breathing I was really scared. About 40 minutes passed before all this craziness subsided. But the next day, the same thing happened, and the next day and so on, Monday through Friday. By Friday I was a wreck.

I kept working, but the first week of December, I called a doctor I had seen a few years ago. (I used to be the kind of person who saw a doctor once every ten years or so.) They had me see the physician's assistant, who checked me over carefully. When I told him the details of when all this had started, his immediate reaction was that I was having problems caused by the renovated workspace. He said I had bronchial asthma and put me on Prednisone for 10 days. Wow! I felt so much better I thought I was cured. How little did I know.

By the time I was off this steroid, I went home to Boston for a couple of weeks for the holidays. After I started working again in mid-January 1994, I was in worse shape than ever. The late-day attacks were getting so severe that my sides were killing me; it felt like my diaphragm was going to explode. Unfortunately, I waited until mid-February to call the doctor again because I was so unaccustomed to calling doctors. I had always been a fit, well person, very athletic. I wasn't accustomed to complaining about my health.

So I called my office to say I was sick on February 14, 1994. Then I called the doctor's office and explained to the nurse that by this point I wasn't recovering at all on weekends from the workday agonies of hacking and wheezing and was barely breathing by the end of each day. I told her that at the moment I was in so much pain that I could only move very slowly. The nurse consulted with the chief doctor and then told me to come in as quickly as I could. She said one of the doctors would wait there for me through lunchtime if necessary.

The internist I saw diagnosed me with occupational asthma. He put me on Prednisone again and tried to convince me to quit my job. He wrote to my employer to ask them to move me to a different location, and he had me document my asthma attacks in great detail. He said the late-day attacks were typical of classic occupational asthma, but he didn't know exactly what substance was causing them.

The only reaction to my internist's diagnosis that I got from my company was a stern warning from my supervisor to keep my mouth shut: "We wouldn't want any other departments here to think there really was something wrong with the environment, now, would we?"

This routine of work and continuing asthma attacks went on and on until May 4, 1994. On that day my supervisor blew up at me and then fired me. I know I was fired because I continued to complain about the workplace environment. That seemed pretty clear because I was the lead person in the unit and was popular with my clients.

After I had been away from my workplace environment for a few weeks, I started feeling pretty good. I hadn't had any asthma problems since being fired. I had, however, had problems a couple of times in a parking lot when it was very hot because, for the first time in my life, I was really aware of the odor of petroleum. When I encountered heavy amounts, it was like being zapped in the brain. Years later I learned that people with MCS call this brain fog.

One day at the end of June 1994, I took a walk around the lake behind my apartment complex. It happened to be a very windy day, and there was a constant cloud of debris in the air from the new apartment buildings that were going up right next to my complex. By the time I got back to my apartment, I wasn't breathing well. Then suddenly my breathing got so bad that I was very frightened. I managed to get to my front door and call to my downstairs neighbor, who was outside. He came up and called an ambulance and stayed with me until it came.

First the fire engine arrived, and they tried to give me oxygen. That didn't help. When a person has an asthma attack, they cannot take in oxygen because they cannot expel the carbon dioxide from their lungs. It makes sense now, but then I was just terrified.

The EMTs arrived pretty quickly. They put me on a nebulizer and took me off in the ambulance to a hospital. This was my first time in an ambulance. First time having an IV stuck into me. First time in the emergency room. First time overnight in the hospital. Lots of firsts. You should see my arms now—we've counted three dozen IV scars.

After my doctor examined me the next morning, I went home on a huge dose of Prednisone. Talk about the jitters—I was so fired up, I couldn't sleep for three days. I redecorated my entire apartment.

Between October 1994 and the fall of 1996, this kind of episode occurred over a dozen times. Twice, both times in March 1995, I went into anaphylactic shock. I didn't even hear the ambulance sirens because I was unconscious. The first time I almost died.

In January 1995, I started seeing an allergist at the university medical center who was supposed to know something about what I had learned was called multiple chemical sensitivity. (I later found out that he considered MCS to be psychosomatic.) He was also supposed to be an asthma expert, but by then I was having other problems in addition to the asthma. My cast iron stomach was betraying me, and I started having episodes of terrible burning in my chest that seemed to exacerbate asthma attacks. Also, for the first time in my life, I had developed really bad sinus problems. My sinuses would sometimes burn, especially when I was in grocery stores, and I started having some yellowish discharge. On top of this, I had a couple of bizarre incidents in grocery stores when my vision would suddenly go. It was as though there were kaleidoscopes in my eyes.

My allergist diagnosed me with gastroesophageal reflux disease and prescribed medications for that. He also started treating my serious sinus infection with antibiotics. It was hard for me to pay for all my medications because I was still on COBRA for health insurance, not yet having been able to find a new job, despite dozens of interviews. I think word had gotten around that I was someone who would react to poor air quality in the workplace and complain about it.

So there I was with all this medication and no job, and very little money available for food. At the end of February, I had more sinus problems. My allergist was out of town, and the doctor covering his patients put me on amoxicillin. Each day that I took this medication my chest tightened up more and more. By the night of March 2, I knew I had to go to the emergency room, so I put on my coat and walked out my front door. Thirty seconds out the door, I turned around and came back inside. I was well schooled in trouble by then. The room started to whirl, but I managed to grab the cordless phone as I hit the floor. Five minutes after the EMTs arrived, I was unconscious.

Later I learned I had stopped breathing, was turning blue, and had almost died. The next day my allergist had me transferred to the university medical center by late afternoon, and I spent two more nights there.

During that same month, I had an anaphylactic reaction to aspirin. I had rarely gotten headaches or other aches and pains before my current health problems started, so I had taken an aspirin maybe half a dozen times per year. What I didn't realize at this point was that my whole body chemistry had changed and I had been damaged by chemicals. So one more round of anaphylaxis, the ambulance, and the emergency room.

But despite all my medical problems, I kept trying to find a job. At last I landed a position as a computer network administrator in the university medical center's operating rooms. I was responsible for 35 operating rooms, their computers, and all the other networked computers in one of the largest trauma operating suites in the area. Many times I would be paged when an emergency arose with a computer in one of the operating rooms.

As I worked in the operating rooms, however, I got sicker and sicker. I had episodes of vomiting and started having daily asthma problems again. Often as I was walking down a hallway in the hospital, it would seem as if the floor was rolling up in front of me. This was more than dizziness, it was really bizarre. Later I learned it was a neurological disorder and was causing me severe vertigo.

Because I had been unemployed for 14 months before I started the operating room job, I had been forced into bankruptcy, although my debts were small. On top of that, my attorneys were busy with a lawsuit against my previous employer. My deposition in August 1996, which lasted eight hours, was an experience I would rather forget. A month later, three days before the jury trial, the judge handed down a summary judgment against me. A year later my appeal was rejected. My former employer was one of the most powerful companies in the state, so perhaps that explains the outcome.

It's hardly surprising I was having horrendous health problems. The exposures I encountered while working at the first company were bad enough. Then when I worked for 18 months in the operating rooms of the university medical center, I headed straight into severe multiple chemical sensitivity. Ask any knowledgeable physician what chemicals you are exposed to in an operating room. The anesthesia is just the bare bones. On top of that, there is heavy use of cleaning products and disinfectants. Nevertheless, within my growing team of physicians at the university medical center, my endocrinologist was the only one who seemed to have a clue about what was causing my problems. He said I should get out of the operating rooms. I took his advice and left the operating room in January 1997, requesting a transfer to a different job in the medical center.

By the end of March, I was rehired by the university to work in its primary center for conducting clinical trials, especially in the area of cardiology, which I thought would be interesting. Everything seemed to be going along pretty well for the first month, until the end of April, when they moved me into another building, which was very smelly. Quite a few people had complained about the air quality there. It didn't take long before I again was having late-day asthma attacks.

On a Friday evening in late May, I arrived home exhausted and did a number of nebulizer treatments for my asthma problems. Then I fell asleep on the couch. About 1 A.M., I woke up on the couch with an awful burning in my chest. By 3 A.M., I was still in a lot of distress, so I went to the emergency room. There they gave me a medication that caused the burning sensation to dissipate pretty quickly. Then they hooked me up to the EKG machine. My EKG was normal. They did some blood work anyhow and kept me lying there for over an hour. Then the resident came in. He told me the blood work didn't look good, and some heart-related enzymes were a bit high. They did some more blood work. An hour later the resident told me I had suffered a heart attack. To say I was in shock is an understatement. They rushed me upstairs to the cardiac unit. I called my parents, who had moved here from Boston in October 1996 because of my chronic illness.

When my internist arrived, he paged my allergist, who soon appeared. They were both shocked that I had had a heart attack because I didn't at all fit the profile for someone who would be likely to have one. No heart disease in my family.

The director of the cardiac unit turned out to be a super doctor, and they did a zillion tests on me in the cardiac unit. After some exploratory poking around, they decided I needed an angioplasty. That was to be done the next morning. But the wild thing is that when the cardiac surgeon went in endoscopically to do the angioplasty, the blockage, or what had looked like a blockage, had disappeared. The surgeon was ecstatic. No angioplasty needed. They concluded that I had experienced a severe cardiovascular spasm. I did not have coronary heart disease, as it is traditionally understood. Nevertheless, the spasm had been potent enough to destroy 25% of my heart muscle, so six months of cardiac rehab ensued.

Finally, at the end of July, I returned to work, and the late afternoon asthma attacks started again. Sometime in August, I had my first follow-up appointment with my cardiologist. I remember lying on his examination table while everything seemed fine. Then within minutes I started getting horrendous pains. Within a couple of minutes I was hooked up again to the machine, and my cardiologist was reading my EKG. He told me I was having an ischemia attack. Lack of oxygen. More heart muscle spasms. I was sent off in the ambulance to a night in the chest pain unit.

My cardiologist made some very astute observations. He noted that the same chemical exposures that seemed to trigger my asthma attacks, which were bronchial spasms, also triggered cardiac muscle spasms. It seems ironic that my work for a division of cardiology exposed me constantly to chemicals that triggered cardiovascular spasms in me.

By the middle of October, as the asthma attacks and the angina attacks continued, my doctors agreed I should not return to that building. I was then fired on the grounds that I could no longer do my job. Finally, in March 1998, under threats of public exposure by my family, my employers backed down on their refusal to allow me to telecommute. I was assigned another position as a technical writer and allowed to work from home. To this day, the entire department admits that no one on the team is more productive than I am.

In the meantime, my sinus disease had worsened considerably. One day I had a severe shooting pain through the right side of my face and raced off yet again to the emergency room, where I had a CT scan. A few days later I had an appointment with an ENT surgeon, who pointed out clearly the bone erosion in my skull and told me that down the road I could be highly susceptible to getting meningitis. That fact alone made me decide to go ahead with the sinus surgery.

I had the surgery in June. I got incredibly well accommodated for an MCS patient at a hospital that doesn't think such an illness exists. (This accommodation resulted because I hooked up with a patient care person who was very interested personally in MCS. She arranged the special room and everything.) At any rate, the surgery was a great success, although it was over three hours long. My surgeon said he had never seen a worse case than mine.

Unfortunately, after the surgery, my MCS symptoms became severe. (They had subsided considerably in the preceding months because I was working at home and avoiding exposure to workplace chemicals.) Although I had been a great athlete and a dancer, now I was falling down everywhere. I didn't trip over things, I just fell, without warning. All of a sudden, I had constant chills, out-of-control asthma, irritable bowel syndrome, a strange metallic taste in my mouth, vertigo, short-term memory loss, and about ten other symptoms.

I finally made an appointment with a physician who is one of the leading researchers in MCS in the country and is particularly interested in reactive airway disease. I have been very impressed with his grasp of chemical sensitivity—he was the one who pointed out that the anesthesia I had been given for the nasal surgery had undoubtedly greatly exacerbated my chemical sensitivity. I also have recently found a local physician who specializes in the practice of environmental medicine, and he too is very knowledgeable about MCS.

Although I am glad to have found physicians who understand chemical injury, my parents and I remain desperate about my future. They are getting on in years and can afford to give me little financial assistance. I don't know how they can stand to see their only child, who had achieved so much, now losing everything. Any chronic illness is bad—no way around that one. But when such debilitating illness results from the ignorance, the complete and utter lack of human caring that my former and current employers displayed, it truly is criminal.

I remember when I met my new attorney in December 1998, I asked him what they pay someone who loses a limb on the job. He told me. Then I asked him what they do for someone with an amputated life.

Sometimes I still just want to slug the people who did this to me. So far, I still have not obtained justice for the wrong, the terrible wrong done to me. I have a badly damaged immune system, as tests have proven. I have organ damage to my heart, gastrointestinal system, respiratory system, and skeletal system. I have permanent vision problems. I still live like a hermit to a large extent—I have no social life except in cyberspace, and after 25 years of hard work I no longer have a career. And I am no kid, being in my early 50s. What can I expect from life at this point? What will I do when my parents are no longer here? These are thoughts that sometimes haunt me and terrify me.

All I can do is try not to allow the terrible rage inside eat me up alive. So what I try to do, sometimes well and sometimes not so well, is to use this rage like kindling to fuel a new direction. It's certainly not a path I planned or dreamed about. Where this effort will lead I don't know. That's the next chapter.

Zach

Eight-Year-Old Child

When Zach was eight months old, we had an oil burner that apparently was faulty, and while we were in the house it blew up. When I looked up, I could see black thick soot coming out of the vents. It was blowing all over, and it covered everything in the house. We moved out of the house, and they did a whole cleanup and servicing of the house because there was this oily black stuff everywhere. At the time, one of the men said to me, "This is very dangerous for an infant."

Later on when we came back into the house, the air was thick with the smell of fumes from the chemicals they had used to clean up the soot. We had thrown out 90% of the things in the house, but the walls had to be washed, and the ceilings were covered in soot also. We did move back in, and that was basically the start of the nightmare.

Within a matter of weeks, Zach, who had been nursed for the last eight months, stopped nursing cold. He wouldn't take food or liquid. He went from 21 lbs. at eight months to 17 lbs. at one year. We took him from one doctor to another. He was labeled "failure to thrive" because he was losing weight and he wouldn't eat. We had to do a lot of things to get nutrition into him. They administered an IV, but as we later discovered he's sensitive to corn, so the dextrose solution bothered him. He was also probably sensitive to the plastic in the IV tubing because plastics are now a problem for him. It was shortly after he had such a bad reaction to this IV that he started patterns of rocking like an autistic person or child does. When he was 13 months old, he would get up on his hands and knees and just bang his head as hard as he possibly could against the sides of his crib until he would either knock himself out or draw blood. Zach didn't sleep, he never slept. He would sleep for 20-minute periods through the night and be awake the rest of the time.

During this period, Zach's doctor suggested we start him on whole milk, and that made things much worse. He became more aggressive, more violent, and kept trying to hurt himself. Milk also caused severe digestive problems. He would be doubled over in agony, clutching his stomach.

When Zach was two years old, we attempted to put him in nursery school. He would only last about 30 minutes before they would call me and say, "We lost him, and we can't bring him back." What that would mean was that he would be kicking and screaming and biting himself on the floor in a rage, crying. That would go on for six, seven hours. Sometimes he would stop for about 20 minutes, but then he would start again. All day and all night, depending on where he went and what he did, this is what would happen.

When Zach was three and a half, we saw the first physical symptoms when you could see on the outside what was happening on the inside. He broke out in open, weeping sores and rashes from head to toe. He would bleed behind his ears and all over his arms. His reactions took the form of looking like burns. They would come out in burn marks and bubble up into these pussy blisters and actually pop.

One time I took Zach to a hospital because I couldn't do anything more for him. I needed help, and I didn't know where to go. They gave him an IV of an antihistimine, which caused a much worse reaction. It took about six or seven doctors and nurses to hold him down and put him into a straitjacket basically. He was out to lunch, gone, his mind was gone. If you looked at him, there was nothing behind his eyes. He was one humongous open sore. He looked like he had been burned. One of the doctors said, "What's going on?" I said, "He's having a reaction to chemicals. He needs oxygen." I had actually brought an oxygen mask and tygon tubing that I kept available because the standard hospital oxygen equipment has strong chemical odors. The doctor said, "It looks like he's been burned." I said, "No, that is a chemical reaction that's happening." They started to put alcohol on his arms that were just all raw. I began screaming, "Don't put alcohol on his arms. You're going to make it worse. You can't give him anything. He needs oxygen." They threw me out of the emergency room and accused me of child abuse.

What happens with Zach's reactions is that though they look like burns, once the reaction is over, the burn mark is gone. Essentially in this case the reaction ended, and I saw it end before they even threw me out of the emergency room because his eyes started to focus. About an hour later the head of immunology came over to talk to me. They could not explain what had happened. Here they had brought in social workers and thrown me out of the emergency room, but he no longer had the burn marks. He did have blisters and healing sores on his body because that is inevitable, but the marks they had seen were gone. So a doctor came over and said to me, "Your son has eczema. He needs a good dermatologist. See ya."

That reaction happened from going to one of those indoor playgrounds, the places where they clean with disinfectants and there are all these plastics and children. I had taken him with the hope, the prayer, he would be all right because he's a child, and all children should be able to play, to do like other children. But in 20 minutes he had open, bleeding sores all over his body.

Zach had another incident where he had an IV because he was running a high fever and was dehydrated since he wasn't eating. As a result of the IV, his whole arm blew up, and the rest of his body blew up. His temperature went up instead of down, and he had a seizure.

Zach has had quite a few seizures. He has periods of seizures from specific exposures to chemicals. Some of them are just basically a blanking, where the doctor describes it as his mind just shutting off for a period of time, and there's no one there. Then he comes back. Other episodes are more like epileptic seizures. He has difficulty breathing, his eyes roll back in his head, and we have to bring him back. Depending on the reaction, depending on what chemical he's been exposed to, his reaction may start with a rage or a tantrum or it may just end in a very severe rage or fit.

The seizures started around age two. It took quite a bit of time to put things together. He had a string of EEGs, and we were able to show that when the EEG is done when he is having a reaction, there are irregularities in his brain. But when it's done when he's not having a reaction, the EEG shows no problems in the brain. To me that is very significant.

From there we went from doctor to doctor to doctor. We went all through Manhattan. We were referred to the best of the best of the best. We saw endocrinologists, pediatricians, allergists. Every kind of doctor there is to see, we saw that kind of doctor, and nobody could really help him. He was diagnosed as an unclassified disorder.

About three years ago, we reached a point where I could do nothing to help Zach. He was in pain all day, 24 hours a day, and he cried nonstop. Finally I found a doctor who suggested that maybe a change of climate and atmosphere to a drier area and a smaller, less toxic community might help. We left Long Island and moved to Arizona. When we got here, for the first time in his life, basically, from when he was a baby until he was five, Zach slept through the night.

Then we had the chore of finding safe housing for Zach because he can't tolerate or is sensitive to all the materials that are used in an ordinary house, like carpet or paint. We have moved four times since we arrived in Arizona. The third place we were in seemed physically OK for Zach, but our neighbors really liked to spray pesticides in mass quantities. One day when it was very, very windy, they sprayed their yard with those big hoses. The applicators were wearing those full-fledged astronaut-looking suits. We went outside before we realized they were out there spraying and got covered, physically drenched with pesticide. I got very sick from the spray, which would happen to anyone, but Zach took a dive. He had been doing really well for some time before this incident, but he took a crash and he did not come out of it. From April until we got him to a special clinic in August, he got worse and worse and worse. We took him to a doctor in Scottsdale who took one look at him and said, "I can't help him. He's going to die." He looked like he was going to die. He weighed only 34 lbs. He had lost 10 lbs. in the course of maybe 8 weeks. He was losing weight drastically, he was emaciated, he had sores all over his body. He could hardly open his eyes because they were so swollen. He had difficulty breathing, he couldn't walk, he couldn't talk. I had to carry him wherever we went. The doctor said, "He needs drastic help," and he ran blood tests on him. The blood tests showed that Zach had tremendously high amounts of various chemicals in his bloodstream like toluene, trimethylbenzene, and benzene. All these different carcinogenic chemicals were just stored in huge quantities in his fat tissue and would be released into his bloodstream periodically. His body was unable to detoxify these chemicals. It just took them in and didn't ever get rid of them.

At the clinic, his doctor put Zach on a rigid three-month long detoxification program to try to get those chemicals out of his body, and it saved his life. He was also put on a very rigid diet because he was sensitive to so many foods. He only ate alternative grains and organic broiled fish, all different varieties of fish, not to be repeated very often because he was on a rotary diet. By the time we left the clinic, he had put on 9 lbs.

Zach continues to improve, but his sensitivity to chemicals has not decreased. The fewer exposures he has, the healthier and happier he is. If you look at him today, you might not know there's anything wrong until he gets an exposure. Just last week we had an incident with one of those blue ice packs that you put in the fridge to get cold and then use them to take away swelling. I had bought one, and Zach was playing with it and dropped it on the floor. It broke and the blue stuff inside sprayed all over his feet. I ran so fast to get that liquid off his feet, somebody would have thought I was crazy. I wiped it off his feet, I soaked his feet in warm water, and I scrubbed his feet, but it was too late. In ten minutes his airway had shut down. His body had swelled up, and he was all red as if we had painted him with red paint. He was tearing chunks of his hair out and screaming in agony as his skin started to burn in reaction to the chemicals. He required oxygen and adrenaline, and a few other injections of things that help when he has a bad reaction.

When I had Zachary, I was 22. I had never had a child. I had never taken care of other people's children. I had no knowledge as to what was normal or not normal or what people consider normal. All I knew was that I had a lot of trouble, and I didn't know what to do about it. Zach didn't sleep, so I didn't sleep. Everyone had an answer for the sleeping problem. Let the baby cry, some said. The baby just kept on crying. My mother used to come over and say, "There's something wrong with your child, but I don't know what's wrong with him."

Here I'm saying, "This happens for six or seven hours a day. This two year old or this three year old screams, kicks, bites, pulls chunks of hair out of his head. He can't eat, he has terrible stomach aches." And no one had any sympathy or empathy. "That's what having kids is like," they would say. I thought maybe that was normal. Then my daughter was born, and she didn't do any of that; she slept almost all day long for the first months of her life. She's been a healthy child and can attend school.

Zachary played Little League this year for the first time. He was really good at it and enjoying it, and to him it was as if he was just like all the other kids. He played the first half of the season, and then I got a call from the school district telling me that they had sprayed pesticides on the baseball field, and it made an abrupt end to his baseball career. He cried for weeks, and he would sleep in his baseball uniform, hugging his baseball bat because he couldn't play. But he is so unbelievably sensitive to chemicals like pesticides that we couldn't take a chance. His doctor thinks that the chemicals could kill him at a level that probably doesn't bother other people, or so they think, so he didn't play the rest of the baseball season.

People tend to be skeptical about Zach's illness and look for other explanations for whatever is wrong with him. After an article about Zach appeared in the local newspaper, some woman went to the school and handed one of the administrators an article on Munchausen by Proxy, the very rare syndrome in which a parent deliberately hurts a child in order to get attention from the medical profession.

Zach is taught through homebound school and is classified under special education as "other health impaired." He cannot attend the school because there are too many chemicals and too many other variables in the school for it to be safe for him to enter the building. When the school doesn't use pesticides on its grounds, Zach can participate in recess or physical education when they are held outdoors.

There have been several teachers who have been alone with Zach when he has had moderate to severe reactions to ordinary things like perfume, scented lotion, eraseable pens, and ink in textbooks. More recently his biggest problem has been that an exposure makes it impossible for him to think. This slowing down of his brain makes it difficult for him to respond rapidly and move away from the offending person or object. Unless I realize quickly what is happening, or whoever is with him does, the exposure is so severe that the reaction can involve a high fever (105 degrees), a seizure, swelling, or vomiting, and it takes weeks for him to recover. I am afraid that one day he just won't recover.

Louise

Business Executive

My sensitivity to pesticides and herbicides began years ago in Hawaii, where they were spraying the sides of the roads with herbicides for weed control. It took me two years to figure out what was going on. I would suddenly have this burning throat, as if I had eaten hot pizza and it had burnt the inside of my mouth. And then I would get a blood blister and it would get bigger and bigger and bigger, and it would pop, and then my mouth would be full of blood. I would also feel completely exhausted.

Finally I realized that I got these reactions when they had sprayed the sides of the roads. I went to a doctor who had been in Vietnam, and he said, "Of course it's herbicides. I was in Vietnam as a doctor. I know." So I was absolutely certain what was happening, and it happened every single time they sprayed. Then I developed a different problem. I was in pain for months and months when I was working on a construction project, and I would go home and take a lovely hot bath and soak in that water. I didn't think anything of it—it was a perfectly normal thing to do. Finally I told my husband, Will, that I was in pain most all of the time, and he said, "Stop taking baths." And I said, "What do you mean, stop taking baths?" and he said, "Because you know there are herbicides in the water." So I did stop taking baths. The next day I just took a quick sponge bath, and within 36 hours I was out of pain, and I had been in pain for almost a year.

At that time I was still very functional. I was working for a corporation, and I was working very hard. About this time Will and I had moved to Santa Fe. We had gone out to lunch one day and were driving back to work when I saw out of the corner of my eye a man who was wearing a protective suit and had on a big gas mask. He had a hose that was six inches in diameter, and he was spraying a tree with pesticides. All of a sudden he went whoosh, and he hit me right in the face as we were driving by with the window open. I said to Will, "Oh, my God, I've been sprayed," and we dashed home. I shampooed my hair twice and threw my clothes in the laundry. I showered and did everything I could to get the pesticide off. I could taste it, so I knew I had taken a little bit internally and I thought I would probably get a headache. As it turned out, I just got sicker and sicker and sicker. Eventually I became almost totally paralyzed. There was a time when I couldn't even open my eyes. I could barely move, but I could still hear. I had to be helped to the bathroom, and Will had to do all the cooking.

One day when things had been so difficult for so long, Will said, "I just can't take it any longer. I'm going to leave." I agreed that we shouldn't let my health problems wreck both our lives and pointed out that we could hire people to take care of me so that he could leave. The next morning, however, Will announced that it was all right, he could handle the situation and he wanted to stay. I'm grateful to have a husband who understands the reality of my chemical sensitivity and has remained with me through all the problems that I have had to face.

I was a professional, and I wanted to keep working, but eventually I just lost my job because I couldn't function anymore. Everything became extremely difficult, and I had at least 30 symptoms. I had a goiter that stuck way out and is still slightly visible. Even though I take thyroid it doesn't go away. I have autoantibodies to my own thyroid. Life has been very hard in the last four years since I was sprayed with pesticide. I come down with symptoms that I didn't even know existed.

It wasn't until we moved out to the country, up in the mountains where nobody sprays, that I began to get better. I became halfway functional again after spending four months here and sleeping outside. About twenty of my symptoms disappeared in the first week. Then I noticed that if I took three steps, my heart would go boom, boom, boom. I realized that I had better stop taking so much thyroid because my thyroid had improved in the clean air. I still take thyroid but not anywhere near as much; I take about 1/30 of the dose that I had taken before I moved to the country.

It's very hard to go from being a very active professional person to being someone who loves the beauty of the mountains but isn't here by choice. I am an environmental refugee, and I think there are a lot of people out there who cannot live where they want to live because of chemical sensitivity. And I have always been a traveler, I have always been a person who loves to stand up in front of a crowd and give a speech and run training sessions. I've trained people from over thirty countries in the world. I've worked for the United Nations, I've written books, I've traveled everywhere. I've dealt with all kinds of people, and I loved it. It was just a wonderful lifestyle.

At VeriFone, where I was the environmental planner, we developed a building that got on the cover of the American Institute of Architects' book on green architecture. There were lots of wonderful things that were happening in my life before I got sick, and to give all that up was extremely difficult, but I had no choice. I couldn't even function as a housewife, let alone a professional. It was quite a major life change that was totally involuntary. I definitely wanted to keep on doing my job. I was getting recognition across the nation from people calling up, saying, "You're doing this? How wonderful, how exciting!" I had contacts with some of the biggest corporations in the United States. It was a job that I definitely wanted and was thriving on and making a difference in the world. It's hard to realize that my life has been sacrificed for a tree.

Of course, after the spraying, we didn't know what I had been sprayed with, so Will did some research and called the company that had done the spraying. They told us it was a combination of two commonly used pesticides, pesticides that you could buy in many stores.

Pesticides continue to be one of my greatest problems. When we were building our nontoxic home, I was living up in a little cabin on the land that we bought, and every morning at 8 o'clock the contractor and his wife would come in, and we would have a meeting. One time they had gone to Santa Fe for the weekend and had come back and on Tuesday morning they came into our little kitchen. Within five minutes I was dizzy and nauseous. I was holding onto the chair and was just about ready to fall onto the floor. They left because I was totally dysfunctional. I had some plastic binders on the table, so I thought, "Oh, well, it's the plastic" because I'm very sensitive to plastic now. So we got rid of the binders. I still felt lousy all day and was nauseous.

The next morning the contractor and his wife came again because we thought my symptoms had been caused by the binders. They sat down, and within five minutes I was just about falling out of the chair, ready to pass out, so they left again. They decided that their shoes must be contaminated and that they needed to wash them, so they thoroughly scrubbed their shoes, but I was still very nauseous and sick. I couldn't eat, and I was just lying in a hammock. I couldn't do anything, even though the construction project was still going on.

Finally on Friday I said to Will, "You know, I think their pesticide is still in the kitchen because I still can't eat." He got a bucket of soap and water and scrubbed down the floors, chairs, and everything, and within a half hour the nausea was gone. It took me all weekend to recuperate, however. I had almost no energy and could do almost nothing, but at least I wasn't nauseous. I could eat again and go down to the construction site. Then on Monday the contractor came over and sat next to me, and Will was across the room. After about five minutes, I heard my husband say, "Oh, my God" because my face turns totally white when I'm in a reaction. My brain stops functioning, and I don't even know that I'm in a reaction. Now all of a sudden, I was leaning to one side because I was going to pass out again.

It turns out that the contractor had gotten pesticides on his jacket. It was wintertime, and he was wearing the jacket in the house and outside, and it wasn't until his jacket had been washed that I could get anywhere near him. I had no idea that anybody could be that sensitive, but if there's any trace of pesticide on anyone, I will pass out and fall on the floor, whether I know they have pesticide on them or not. I don't even smell it, so I don't even know that it's there.

I think one of the greatest hardships of being isolated and having to stay in one place and not be able to go to other places is that you lose your family ties. My mother had Alzheimer's disease, and I was always very close to her. We loved each other dearly, but I had to have Will take care of her rather than me. He was the one going to visit her. She got so confused at the end that she thought Will was her son, and she didn't remember that she had a daughter because she hadn't seen me in a long time. When she died, I couldn't even go and watch the dispersal of her ashes or do any grieving with the family, even though I loved her dearly.

Will, Louise's Husband

When Louise was sprayed with pesticides in Santa Fe, she and I were working for VeriFone, which is a company we helped found in 1982. At the time she was sprayed, we had about 70% of the world's credit card transactions going through our equipment. As we were founding VeriFone, Louise and I traveled pretty much all around the United States to the various different VeriFone operations. She became involved in the company after a couple of years because she figured out that if she ever wanted to see me again, she would have to be part of VeriFone. And she was a very active part of VeriFone. She was responsible for our training operations and then later on she became our environmental planner.

Prior to that, Louise and I had worked together in the United Nations. We were both UN officials responsible for a number of different development programs throughout South and Southeast Asia. After we came back to the United States, she was also heavily involved at the University of Hawaii assisting in the production of educational materials in Micronesia in twelve different Micronesian languages.

So Louise was a very active professional before she was sprayed with pesticides. After she was sprayed with pesticides, she was only a very small shell of her former professional self. Even now, many years later, even on very good days, she can do only about a third of what she was able to do before she was sprayed with pesticides.

I know that the dichotomy that a lot of people make between the environment and business doesn't hold water because of an experiment we did in Costa Mesa, California. We were growing our business down there. We had an 80,000 square foot building that was a mixed manufacturing facility and office space, and we needed to expand because we were growing very fast. We had a blank building next door that was a concrete tilt-up shell. It was also about 80,000 square feet. Louise, who was our environmental planner at the time, came up with a plan and said to us, "Now look, if we do this new building differently, we can improve the health of our employees and we can also help finance this by saving energy and saving money." She put together a plan and presented it to our senior management team, and the senior management team bought into the plan. The result was another 80,0000 square foot facility, but this one was very different from the original. It was different in that it didn't use toxic chemicals in the carpets, in the furniture, in the paints. We had a pest control policy that didn't allow pesticides to be used in the building. We were using natural barrier techniques and other kinds of techniques like boric acid, and the result really was gratifying. We saw a 40% reduction in absenteeism between the two buildings, and we had roughly the same employee population in both. They were about the same size building, and they had approximately equal functionality. When we went to the employees and said, "How do you feel?" the employees in the new building said, "You know, I don't have that headache that I used to have at 3 o'clock in the afternoon. I'm not dragging at the end of the week the way I used to. In the old building, if they broke out a can of paint and started trying to paint, I would usually get sick and have to go home. In the new building, the paints that you're using are nontoxic, and they don't bother me anymore." So our experience was that by constructing a building in a nontoxic way, we were able to dramatically enhance the health and well-being of our employees, and as we look at it, the single most important competitive weapon of the twenty-first century may very well be having healthy employees.

Unfortunately, Louise had just finished this building when she was sprayed with pesticide. She got so sick so quickly that she wasn't able to go out to the building's dedication. I had to give her speech, and when I came back from Costa Mesa, she had deteriorated so badly that she was just barely conscious most of the time.

Jim

Chaplain

Before I developed MCS, I served as the chief chaplain in a VA hospital in South Dakota. I loved my work and frequently worked 12 to 14 hours a day. When I went to see my doctor for my annual checkup, he would usually say, "Well, Jim, you are disgustingly healthy as usual." I jogged, hiked, skied cross-country, and was not allergic to anything. Then in January 1993, I was moved to a renovated office containing new carpeting. I got very sick within the next few weeks. I felt like I was drunk—I would slur words, I was unable to think, and I became uncoordinated. Later that spring I saw a neurologist who diagnosed me with toxic encephalopathy caused by a toxic exposure. He stated that I had severe damage to my brain and central nervous system and other organ systems. In May, I resigned from my position at the urging of my neurologist, who was convinced that my illness had been caused by exposure to the chemicals in the remodeled office containing new carpet.

As the days passed, I realized that I could no longer tolerate being around lawn sprays, petrochemicals, perfumes, new clothing, hair-care products, or even printed materials. When exposed to these things, I would get symptoms like the new carpet had given me; I could not think clearly and my speech became slow and slurred. My body just would not function properly. Focal brain seizures caused my body to jerk at night, awakening me from a sound sleep.

During the first few months of my illness, I was forced to live outdoors much of the time because of the herbicides and insecticides that are so frequently sprayed on lawns in our small town. I was forced to retreat to the hills, where I slept in our station wagon. I would return almost daily, wearing my respirator, to shower and pick up food prepared by my family. Then I would head back to the hills. Most of the time I ate cold food. I was just beginning to realize I needed all organic foods.

In August 1993, my wife, Jan, and I bought a 1966 Airstream trailer and took it to a higher area in the Black Hills where some friends of ours owned some land. There we began a refurbishing job on the trailer. While we were doing the work, we slept many nights in our old van with the doors open for maximum ventilation.

In the late fall, we headed for the desert Southwest in an attempt to find cleaner air and a safer environment. In our travels we visited many locations in New Mexico and Arizona. We met many wonderful people who also suffered from environmental illness. We learned a great deal from these people, who are now among our treasured friends.

Unfortunately, much of the time that we spent in the desert, I was weak, tired, and struggling with increased symptoms. Irrigation projects on the Colorado River have made it feasible to grow crops in more regions of the desert, so there is now crop spraying for at least seven months out of the year in these areas.

After we had spent five months in the desert with no improvement in my health, we realized that the Southwest would not be the answer for us. We were grateful for the friends we had made, but we realized our support system was in South Dakota. We missed the love and support of our family as well as good friends.

We returned to the Black Hills in a heavy spring snowstorm. To my amazement, I found I had much more energy in this environment than in the desert. I could walk in the snow and keep up with Jan, whereas I had been dragging behind her when we walked in the desert. I even had enough energy and stamina to shovel snow. The air was clean, perhaps cleaner than the air in the Southwest.

Following the advice of my physician, we made plans to move to a higher and more isolated area of the Black Hills. We sold our home and purchased six acres in the ponderosa pines where there were few neighbors. On this land, with the help of many supportive friends and family, we built the first modern straw-bale home in the state. It is built of organic rye straw and the exterior is plastered with portland cement-stucco. There is no paint on the inside except for the doors. The interior walls and ceiling are sprayed with white portland cement. The radiant slab floor is totally covered by ceramic tile.

Jan, who is a nurse, began working full-time, and I attempted to serve as the contractor for the house. This was a major mistake because I tried to do much of the construction myself. By being exposed to the products in our home before they had aired out sufficiently, I sensitized to them and began to react to them. As a result, I could not live in our home for over a year. Before I could return to live in the house, Jan and a crew of workers sprayed the interior walls and ceilings with several coats of a safe sealer. I then stayed out of the home for four additional months while the interior surfaces cured. Now I can tolerate our "safe home" very well. It looks a bit like a gnome home, but to me it's a palace.

We now have many inquiries about the house because increasing numbers of people are considering the use of straw. We no longer hear the jokes about the three little pigs. Instead folks are realizing straw is a good insulation with an R-rating of 40+. Building a straw-bale house is a labor-intensive project, however, and requires persistence. Many people ask us about the fire danger, bug infestations, and mold risk in straw-bale houses, but if they are built correctly, none of these potential problems arise.

In spite of our drastic change in lifestyle and building of a safe haven, my illness did worsen. Most debilitating was the progression of the body jerks to full-blown seizures of the grand-mal type. In April 1997, I had further medical tests that indicated I would be a good candidate to try a new anticonvulsant, and it did help me greatly. It allowed me to sleep and regain tolerance for more chemical exposures. This tolerance has allowed me to do a few more "normal" activities. The medication does have side effects and must be monitored carefully. I now have been able to reduce the dosage some as I find magnesium helps control my seizures.

Throughout my struggle with MCS, my loving family and wonderful friends have been understanding and supportive even when they could not totally understand the illness. My wife, Jan, has helped our family and friends understand the illness. She researched the condition and shared educational tapes and articles with them, which has made a tremendous difference. Let's face it—this illness is not easy to understand. Jan's sister put it best when she said, "I believe what you're telling me about Jim's illness, but it sounds like something from the *Twilight Zone* or outer space."

When I was first damaged by a toxic exposure, Jan and I had no knowledge of the illness. We had never heard of multiple chemical sensitivity, nor did we know of anyone else with the condition. As we began to understand the complexity of MCS, we started to realize the scope of the changes we needed to make, but we still had many unanswered questions. Where could I safely live? What should I eat or avoid eating? How could I make clothing safe for me to wear? How could I avoid further toxic exposures? We didn't know where we could go for help.

It was during this time of unanswered questions and my struggle with isolation and depression that God gave us a vision of how we could use our difficulties to help others who experience the same problems. Out of these plans would grow the Hand of Hope Ministry. We wanted to assist individuals to regain their emotional balance while learning to cope with the many challenges of MCS. We wanted to help these people assess their environment and learn how to make it less toxic so that their health would improve. We hoped ultimately to build a network of support that would allow others to find a way out of isolation, loneliness, and depression about their situation.

Jan and I are excited and hopeful about the future. We recognize that at this point MCS is a chronic illness, but we also know there is much that can be done to empower the person who is chemically sensitive. The first goal is to accept the fact that you have MCS and the changes it brings to your life.

I remember vividly when I reached this point. Prior to that time, I was angry. I wanted to continue my hospital chaplaincy, and I could not do it. I felt hurt and cheated. We had been camping in our Airstream trailer on a friend's land in Arizona. Nearly every night our neighbors built a huge bonfire and sang songs. It looked like fun, but the smoke caused me to have seizures. We had to flee to the mountains to sleep. Frequently at night I would have dreams about my work as a chaplain. The dreams made me feel good, useful, and needed.

When I would awaken in the morning, I was hit again by reality. My life's work had been pulled away from me at age 55. I was depressed, and I'm sure I was difficult to live with much of the time. Then one night, lying in our old GMC van out in the mountains of the desert, I prayed that God would help me accept the illness. I prayed that he would help me take control and make the best of my life as it was. When I awoke the next morning, instead of feeling empty and angry, I was filled with a sense of peace and acceptance. Life would be good again.

That was a major turning point for me. Of course I still missed my old life and my ministry, but now I had strength to move on. I knew God had a new plan for our lives. I clung to God's promise in Romans 8:28, where it says, "We know that in all things God works for good for those who love Him." I knew God did not cause my illness, but he would use it for good.

In the following years, we began to take control of the illness. We made our safe house in a region that was healthier for me. Then I began to learn how best to care for myself. I learned that no one could know my body and mind better than I did. We found competent alternative sources of medical care. In addition, by pursuing a nontoxic lifestyle and avoiding exposures to toxins, I have regained some tolerance to the exposures that are unavoidable.

Over these past six years, however, I have found that it is very easy to slip into the role of victim. This was especially true in the early years before I had knowledge of therapies that have helped me maintain physical and emotional balance. One such experience stands out in my mind. One day we returned to our new straw-bale home after a month's absence. This was before we had sealed the walls and ceiling with the safe sealer. I was hoping that my sensitivities to the inside stucco had subsided, but after I had been in the house for about five minutes, I flew into a rage. There was no reason for the rage, and this was totally unlike me. I smashed the cup from which I was drinking on the tile floor. I stomped out of the house into the cold winter air. In ten minutes or so, my head had cleared from the chemical reaction.

I realized that my rage had been caused by the chemicals on the walls and was not a psychologically based reaction. However, I started questioning God again. "Why me?" "How could you let this happen?" It took weeks for me to pull myself away from focusing on these negative thoughts, but with God's help, I eventually did. I knew that if I was to move from victim to victor, I had to get beyond the "poor me" attitude.

In my work as a hospital chaplain, I had watched too many people with chronic illnesses become bitter. They made life miserable for the staff and for their spouses and families. I knew from this experience that I had a choice to make. Would I become bitter or better? Would I become self-centered and dwell on my own illness like many patients with whom I had worked? Or would I try to look beyond myself and try to help others in need?

I decided to seek what good could come out of my illness to benefit others. Slowly peace returned. I trusted God and knew he would show us the way. Through Hand of Hope Ministry, Jan and I have watched good growing out of our suffering. That has brought meaning back to our lives. It has given us strength to keep on going. Once again I feel productive and useful. God has blessed us each step of the way, and I know he will continue to do so.

Julia

Gulf War Veteran

From December 1969 to December 1970, I was a Navy nurse stationed on a hospital ship in Vietnam. When we were not on the ship, we would sometimes go into areas that had been defoliated by Agent Orange. When I returned from Vietnam, I began having irritable bowel syndrome and lactose intolerance, although I had not had these problems previously.

In the Gulf War, I served as a commander in a fleet hospital, which is a tent hospital that holds 500 beds and has 1000 personnel assigned to it. Even before the rest of our group arrived in the Persian Gulf, the Navy Seabees had gone there to pour concrete pads to set up our tents for the hospital. The area where the hospital was set up was in a region of the desert that was formerly an industrial park. Five miles away was a fertilizer plant. In late January, the Seabees who were pouring the concrete heard a loud explosion near them. Some of the Seabees' shirts turned purple and many had burning sensations on their hands and face. Some wondered if a scud missile had hit the fertilizer plant, releasing some chemical into the air. All the Seabees were told was that the chemical was probably an industrial solvent. Many of those Seabees are really sick now, and when I first arrived at the compound, I noticed lots of dead rodents around the area. I also noticed that there were no insects around.

Once the concrete pads were ready, we came in to set up the tents and the sleeping quarters. I started having open sores on my right foot right away after I arrived—they were located near some ventilation holes in my boot.

Then in February there was another explosion, and immediately after the explosion an ammonia smell came through the camp. We put on gas masks, but the hospital air conditioning and heat pump system still brought in outside air. We didn't have any chemical alarm system in our camp, but we were told that there was nothing to worry about. Soon after this explosion, however, I started having a rapid heart rate and uncontrollable blood pressure problems.

We began noticing some odd things around the hospital. Some of the synthetic fabrics on the cots were beginning to disintegrate. A lot of soldiers were coming in with uniforms covered with oil, and the fabric on these uniforms was beginning to fall apart. We didn't know whether the oil was destroying them or whether there was some other chemical attacking them. The oil on the uniforms probably came from the oil well fires that were burning all over the area. When it rained, the rain would be oily. It left an oily coating on everything, and it was almost impossible to clean that off the tents.

When I left the Persian Gulf, I was ill. I was having fevers and had become sensitive to many chemicals. I can't stand to go into any kind of store that sells fertilizer or pesticide. We don't use anything like that on our lawn anymore. I use all natural fibers in my house, and we have taken out the carpets and replaced them by wood floors. I can't wear polyester clothes now because I start swelling if I do, but before I went to the Gulf War, I had no problem wearing polyester.

Rashes are a problem for me now, as for almost all the vets with Gulf War syndrome. The rashes are like small pustules that leave scars. My joints swell up all the time; I participated in a study at the University of Pennsylvania that showed my joints were all swollen and that I have bursitis and osteoarthritis. The arthritis in my hands is really severe now, particularly in my thumbs and wrists, which are enlarged and constantly inflamed. I've had severe headaches and chronic sinusitis since I returned from the Persian Gulf. I've had my gall bladder removed and a foot of my colon removed because of chronic diverticulitis (a condition I didn't have before the Gulf War). My ovaries and uterus were removed because they were in a precancerous state. I've had stomach polyps and rectal polyps, and I have a slight case of Barrett's esophagitis, which is a precancerous condition of the esophagus that you have to keep an eye on. I have some brain lesions, and I've also developed diabetes this last year.

The Veterans Administration still doesn't recognize that my autonomic nervous system has been damaged as a result of my service in the Gulf (causing the rapid heart rate and uncontrollable blood pressure fluctuations). When I went to the Navy for those problems, they said it was stress. Fortunately, the National Institute of Health was doing a study on stress and the effects of stress, and they sent me to participate. The researchers there said that I had autonomic nervous system damage—cause unknown. It's hardly surprising that I was the last Gulf War vet they sent over to participate in that study.

After many years of battling, I have finally received disability status from the Navy on the basis of autonomic nervous system damage, but they won't recognize the cause. I'm also listed as having chronic fatigue syndrome and post traumatic stress disorder.

I would still like to know what chemicals I was exposed to at our fleet hospital site in the Gulf. There must have been some reason why those Seabees' shirts turned purple. We've asked the government to analyze those cement pads that the Seabees were pouring when the missile exploded, but it's never been done.

After all these years, the Veterans Administration and Department of Defense still do not recognize Gulf War syndrome as an illness with a CDC diagnostic code. If there is no illness, then of course there cannot be any cause or responsibility. The DOD has admitted there were some exposures to "industrial pollutants" and chemical weapons, but the latest "studies" by the Navy have shown no side effects from chemical exposures. These studies were done in military hospitals, however, not VA hospitals, where those soldiers whose illness is so advanced that they have had to leave the service would be seen. Most of the time, low-level chemical exposure does not cause an immediate effect but can cause long-term damage. I'm convinced that the government does know the effects of low-level chemical exposure, be it Agent Orange or Persian Gulf exposures, and yet does not accept responsibility for its part in the medical problems. If they don't accept responsibility, then there is no hope of protecting future military forces or cleaning up our environment.

Diane

Registered Nurse

When I was diagnosed with multiple chemical sensitivity in 1993, I had never heard of the illness. Although I had been a registered nurse for twenty-seven years in one of the best teaching hospitals in the world and was a graduate of its school of nursing, I had no information on MCS, nor did the hospital library.

When I was a baby, I had eczema that was attributed to carrots. Unfortunately, I also happened to live only a mile from what is now a Superfund cleanup site. By the age of nine, I was having unexplained fevers, abdominal pain, and failure to grow. This condition was treated for several years, and I was hospitalized twice, without a diagnosis. Then my health stabilized until 1964, my second year in nursing school, when the fevers and abdominal pain reappeared. I was hospitalized again for tests, which were all normal. The diagnosis was stress and the vivid imagination of a student nurse. I continued to deteriorate, however, until January 1965, when I had emergency surgery for a perforated intestine and peritonitis. Now the diagnosis was Crohn's disease. At that time, Crohn's was considered a terminal illness, but that's no longer true. Fortunately, I remained in remission and continued an active life without many changes in goals, as long as my diet was fairly bland. After graduating from nursing school, I became a staff nurse at the same hospital. My doctor limited my work schedule to a 32-hour week when I was doing bedside nursing.

Then in the winter of 1969, my health failed again, with a series of added symptoms: dizziness, nausea, problems with balance, joint pain, weakness (combing my hair was a workout), anxiety attacks, exhaustion, abdominal pain, and ophthalmic migraines. I was hospitalized for a week, but all tests were negative. The diagnosis was again stress, and the treatment was rest and referral to a psychiatrist. About three months later, I went back to work part-time. By then I had become very sensitive to medications and could usually not tolerate a child's dose without side effects.

By the mid-1970s, I was working in an intensive care unit. Treating patients in an ICU can require daily contact with a wide variety of chemicals in disinfectants, pharmaceuticals, gloves, deodorizers, thermal printers, vinyl

products, plastic packaging, and anesthetics offgassing from patients returning directly from the operating room. As with all round-the-clock occupations, cleaning, painting, building repairs, and floor stripping went on at all
hours, seldom with any regard for air quality. The combination of all these
products created a chemical soup.

In 1982, I developed a cough one day after mowing the lawn. (Was it
the fumes from the mower, grass terpenes, pesticide on the lawn, or just
pollen?) In the next few weeks, I began having very frightening coughing
events at night during which I could not get enough breath. One night these
episodes became so bad that my husband took me to the emergency room.
The diagnosis was once again stress, and jogging was recommended. The
next day I saw my regular physician, who told me it was a virus that would
last about six months, and he gave me a drug for the bronchospasm.

In 1985, my neurological symptoms of loss of balance, sudden falling,
dropping things, dizziness, and joint pain returned. I was referred to another
neurologist, who found little on formal tests but did find a drug that I could
tolerate that reversed most of the symptoms. Also at this time my husband
heard an NPR radio program on chlordane and wondered if a contributing
factor to my deteriorating neurological condition might be a professional
chlordane termite treatment we had done in our home in 1981. When we
had the house tested for chlordane and inspected by the state, we learned that
the chlordane application had been illegal and that the levels in our house
were way above EPA limits. Despite some talk by the state of condemning
our house, we moved out for two years while it was "cleaned up," including
replacing basement sills, adding four inches of concrete to the basement
floor, and throwing out much of our stuff. Physicians dismissed chlordane
as a cause for my increased health problems, however. (Chlordane was taken
off the market years ago.)

By 1989 a pattern of deteriorating health among our hospital staff was
becoming increasingly apparent. Staff members with no previous history of
allergies were becoming steroid-dependent asthmatics. Exhaustion,
dermatitis, and severe respiratory and sinus infections were common. Four
out of five head nurses in my unit had resigned with increasing new health
problems. Non-nursing staff also had ongoing medical problems, either
respiratory or neurological. Most staff who wore contact lenses became
unable to tolerate them at work because of eye irritation. Morale was down
and irritability was high.

One day another nurse who was having an asthma attack asked me how
to file a complaint with OSHA. I suggested that we exhaust all the
appropriate channels before going to OSHA, and I will always regret that
response. The next four years of exhausting appropriate channels was a

complete waste of time, and many staff suffered as a result. My alma mater, my second home, the hospital that had saved my life when I had a perforated intestine, became a place that made me ashamed to be a nurse. Sadly, I believe, most hospitals were treating staff the same way. We did not have a union. Patients received the best care possible, but many staff were getting sick. We blamed symptoms on any old thing and tried to cope. We had no education on sick building syndrome, occupational asthma, or MCS.

There are so many memories of those last four years of my career as a nurse that I'd like to forget. Diesel fumes from "reversing the generators," floor stripping, painting, and the use of a particular EEG paste seemed to make many staff members sick. When our unit was painted in 1989, the painters brought in fans that vented the odors while they worked, but they took the fans with them at the end of their day. That left the hospital workers to inhale the strong fumes of drying paint. We tried to get fans, without success.

I documented the painting incident in an official report and wrote a letter to my head nurse and told her I was thinking of going to OSHA. There was no response. We contacted Employee Health for help, and eventually they did conduct a survey of health problems among the staff. Employee Safety was next, and after endless complaints and documentation of air quality problems, inspectors actually crawled into the vent systems in our corridors. I happened to be working during this inspection; when the inspector descended from the system, the look of horror on his face said everything. About 1992 the Employee Health nurse did meet with the staff and told us that the vent systems were not functioning properly, but there was no money to fix them at that time. Also, she said the vents would work better if windows remained closed at all times. To ensure this was the case, all windows were bolted shut about two months before I left.

Eventually I went to the local OSHA office to find out how we could learn what was in various cleaners and other problem substances. I was instructed to contact Employee Safety for the material safety data sheets (MSDS). I had never heard of these product information sheets, which were required by law to be in easy access for every product in use. Employee Safety told me they could not possibly provide these documents for everything we worked with, but allowed me to select three. I chose an EEG paste, a new disinfectant, and acetone, which we were using to wash the paste out of a patient's hair. The EEG paste was stored in a laboratory adjacent to our unit. I went to confirm the name of this paste, only to find the staff there were also sick. Within forty-eight hours of my requesting an MSDS for this paste, they installed a ventilation system in that lab. It was weeks before I received the requested information sheets that were supposed to be kept readily available in the hospital.

In the meantime, staff were becoming sicker. In a two-year period, there were six pregnancies among our staff. One woman had a stillborn baby, three had miscarriages, and two had preeclampsia. (This is a potentially life-threatening condition of late pregnancy.) One study showed a fourfold risk of preeclampsia in women exposed to solvents (Schettler et al., *Generations at Risk* [MIT Press, 1999, p. 79].) The sixth mother was our first case of serious latex allergy, which required special precautions for the birth. She is now totally disabled with latex allergy—another devastating illness, often accompanied by MCS, which the medical profession is only now beginning to grasp. The mother of the stillborn infant subsequently had two healthy babies, but not without complications during the pregnancies. Tragically, she died of cancer when the oldest child was barely two years old.

At this time, I didn't realize how much the chemicals in the hospital were affecting my own health, and I attributed most of my symptoms to stress or my Crohn's disease. I did not grasp that my chemical sensitivity to a few things at the hospital could develop into multiple chemical sensitivity that would forever change my quality of life.

Exhaustion had forced me to cut back further to a twenty-hour week, the minimum for benefits. My days off were spent recovering from days at work. What was totally new was that I couldn't understand why I could smell things so strongly that I had scarcely noticed before. Fragrance left me not only in pain, but confused. I was reversing numbers, and math was becoming very difficult. I could no longer take notes—I could hear the words clearly, but I couldn't figure out how to write them. I had frequent chest pain, irregular heart beats, and numbness of my face and extremities. I fell frequently, my foot dragged at times, and often I was so weak I could barely climb one flight of stairs. At times I had great difficulty finding the words needed to convey specific information and often couldn't remember even my husband's name.

A friend suggested I make an appointment with her boss, a physician in occupational medicine where I worked. He diagnosed me with MCS, referred me to a support group for people with MCS and, backed by Employee Health, said there was no safe place for me to work at this hospital. I soon learned this would be true for any hospital. My nurse manager said it wasn't convenient for me to go on a leave of absence for several more weeks, so I continued to work. I still didn't get it.

After the diagnosis, some of my colleagues apparently decided to test out the reality of this illness they had never heard of and clearly considered psychological. I was having a terrible time with fragrance, and on two occasions the nurses to whom I was giving the off-duty report got out a bottle of perfume, opened it, and daubed themselves as I struggled to relay information. These chemical exposures caused me to lose my ability to organize my thoughts, which was important when I was relaying crucial information about patients in an intensive care unit. The second time this happened, an aide and young secretary who saw what was happening escorted me to an area where a window could be opened for me to get some air. I never again gave a report to certain nurses in person; I found excuses to call it in from other wards. When I began my shift, I would have to remove fragrance strips from my patients' rooms, and I was constantly dodging the spraying of air fresheners, which were frequently used on the ward. The buffer machine also sprayed strong cleaners on the floor on a daily basis.

And still I kept working, but reality was beginning to sink in and I knew I should get out of there. The nurse in charge of Employee Health said, however, that my doctor's letter wasn't clear enough for a leave of absence. At that point, my physician wrote a one-line letter announcing that my leave of absence would begin at the end of the next day. When I met with the Employee Health nurse on that day, she was wearing a lot of fragrance. When I mentioned it was making me ill, she replied, "In this day and age people do not feel clean without fragrance; you will just have to get used to it." She also claimed I was the only person she knew with MCS. The last I heard she had been promoted and was testifying against nurses with MCS. Whatever happened to our 1960s teaching that prohibited nurses from wearing perfume on the grounds it might make patients sick?

That Friday I worked one last shift, said good-bye to the staff on duty, and walked down the corridor to the elevator. As I entered the elevator, a nurse called down the corridor to me, "Help us!"

I had always wanted to be a nurse and had no interest in leaving bedside nursing. God, how I miss the patients! For over a year, not a day went by that I was not at least briefly in tears because I had had to leave nursing. But over that first weekend after I left my job, I could think only about the staff that I had left behind, especially those who were sick and those who had died.

As with others with MCS, my life has changed dramatically. I had to throw out a bunch of stuff like carpet from my house, I had to change cleaning products, get air and water filter systems, buy clothing of specific fabrics, and eat organic food—the list of changes seemed just endless.

I often compare my MCS to my Crohn's disease: both are chronic, have no known cause, have no cure, are not usually life threatening, are difficult to diagnose, fluctuate from day to day, and require lifestyle changes. The two major differences: with Crohn's, although I had to live with the illness and change my diet, I still could live in my house and get out in the community and work. With MCS, most of the lifestyle changes I need to make must also be made by others as well if I am going to be able to be with them. This is an extraordinary request to make of family and friends. Crohn's is accepted by the AMA, while MCS remains in limbo, a rejection that is totally incomprehensible to the people with this illness.

I have been extremely fortunate that members of my immediate family have changed their lifestyles to accommodate my condition enough so I can live with them. Others modify their lifestyles when I visit. They should be healthier because of the changes and I may be easier to live with when I'm not reacting to something, but there are very few illnesses that necessitate a companion to give up so many personal items. I've been closely involved with support groups that connect me to others with the same MCS limitations, including former medical workers. I keep hoping the world will change to allow access to it again, but at the moment I can only try to cope creatively.

Robert, Diane's Husband

Before MCS: live theater, fine hotels, and flights to Europe. After MCS: television, tenting, and trips only by car, to escape perfumes, potpourri, and airplane petroleum and pesticides. Where once we spent mere hours in the air, we now spend days on the road to avoid the chemical soup that passes for air in planes and rental cars. When campgrounds are unavailable or too smoky and motels too fragranced, the back of our station wagon becomes the lodging of last resort.

Before MCS, home life included candlelight dining, evenings by the fire, wall-to-wall carpeting, and a soft mattress of foam and polyester paired with a box spring clothed in fabric. Now the candles are gone, steel foil covers the fireplace opening, all floors are bare wood or tile, and our mattress is made of organic cotton batting and organic cotton fabric above a set of bare springs. The window, once open to air thought clean, now seals out fertilizers and pesticides from the neighbor's lawn and garden. Our lullaby, once the chirp of crickets, now is the hum of an air purifier.

Before MCS, when I returned from trips in my dry-clean-only suit, Diane would be there to welcome me home in a stylish dress in artistic colors. Now, the dress is off-white, beige, or olive (made of undyed organic cotton), and Diane has to retreat to safety until my every item of clothing has had many lengthy washings in baking soda and fragrance-free detergent to reduce the fragrances, air fresheners, formaldehyde, fuel residues, and other fumes that I've picked up in hotels and airplanes. Even the birthday suit itself requires several soakings in the shower.

Before MCS, gasoline powered our mower, leaf blower, and snow blower. Now electricity cuts our grass and blows our snow; our leaves we rake by hand.

Before MCS, fruits and vegetables came from the full palette offered by the largest and most convenient supermarkets. Now, we buy organic produce only, from the more limited selection offered by smaller specialty markets a dozen miles away.

Before MCS, both of us worked at professional occupations, Diane as a nurse, I as a physicist. Now society receives only half the benefit from what, through years of education and experience, it trained us to do. Even my contribution is less than it might otherwise be, what with all the distractions of caring for someone who is disabled. I have had to abandon my dream of working for a year or two in a lab in Japan or Europe.

Why would a person not afflicted with MCS consent to change his or her lifestyle to this degree? At times the temptation becomes almost irresistible to buy into the belief system sold to a gullible public by the chemical industry and the medical establishment: "It's all in your head!" At times one lashes out at another's need for constant care. Time and again one explodes in frustration at being so powerless to help. However, even more powerful and enduring is the desire to provide the best of care for a loved one in need. That no one yet has found a cause for MCS merely suggests that no one yet has looked for the right evidence, using the right tools. If one looked only for bacteria, one might say that the flu, too, is "all in the head." Does the absence of bacteria and viruses mean that cancer is "all in the head"?

One can't deny that the new lifestyle has its rewards, too. It's healthier; all those fragrances and fumes can't be doing even a robust person any good. The electric mower, with its heavy battery, is harder to push, but how pleasant it is to escape the fumes and noise of its gasoline sibling. The snowblower, outlet powered, is cleaner, quieter, and lighter as well. Organic fruits, vegetables, and dairy products are more expensive and take more effort to obtain, but are tastier—sometimes markedly so. Beyond this, there are the rewards of being needed, caring for someone special. The greater the external opposition, the greater the internal bonds. What stronger bond than the one between a couple for whom it's "you and me against all odds"?

James

Air Force Sergeant

I have Gulf War syndrome, but I never went to the Persian Gulf. I did have the anthrax vaccine, however, and I also took the pyridostigmine bromide (PB) pills that the Gulf War soldiers were given.

When the Gulf War broke out, I was in a mobile unit that was being sent to Germany as part of an antiterrorist team. I knew all my shots were updated because I had just got them the week before, but the sergeant in charge told me I was on the list to get some more shots. I believe this mistake occurred because I had been on a list to go to Saudi Arabia, but I hadn't been home for the deployment.

A day after I received the anthrax vaccine I felt like I had the worse case of flu that I had ever had. I hurt all over and had a headache from hell. A day later I developed a strange rash. When we arrived in Germany, I was the only one given a packet of pyridostigmine bromide pills, apparently because my name was still on a list for deployment to the Persian Gulf. At any rate, I took all of them.

When my unit returned from Germany, I felt like I had the flu. I was tired all the time, and I knew something was very wrong when my health didn't improve as the weeks passed. At this time, I had the highest priority clearance to guard aircraft. Every 15 minutes or so, I had to walk around the airplane to be sure everything was OK. In between those patrols, I would sit in a tiny little shack that was heated by a diesel ground heater. The fumes were horrible, and I started getting terrible headaches every day after work. I was also having really bad diarrhea, and that has continued to this day. My joints and muscles ached, and I began having memory problems. Once I left my M-16 rifle and a full clip at a snack shop. My memory is so bad now that I can't even drive on back roads here where I've driven since I was 16 because I forget which way to go.

I also became very sensitive to various chemicals. I couldn't stand being around cleaning chemicals, and diesel fumes or exhaust always gave me horrible headaches. Finally I just couldn't keep working around all the diesel exposures on the flight line, so I left the Air Force. For a couple of months, I drove a truck for a man who was laying cable for TV's, and that worked out only because he told me where to go and when to turn. The worst thing was that my diarrhea was so bad that we would have to stop every half hour for me to use a bathroom. Of course, after a while I had to quit that job.

By this point, all my symptoms were intensifying, and I couldn't see how I could support my wife and two little children. The Veterans Administration refused to recognize that I had any real health problems. They kept telling me I was suffering from stress, and they put me on Prozac. Finally they gave me $119 a month for 20% disability related to an old back injury I got the first year I was in the Air Force. Of course, the four of us couldn't live on $119 a month, so we had to give up and go on welfare. After three more years of fighting the VA, I finally obtained 100% disability for the back injury because they concluded I could no longer work.

At one point I was hospitalized for a shigella infection, and after that stay in the hospital with all the exposures to disinfectants and strong cleaners, my multiple chemical sensitivity became really intense. Exposures to things like strong cologne or strong cleaners make my eyes water and my nose clog up. I also get a severe headache. Like many Gulf War vets, I'm plagued by night sweats. Soon after the hospital stay I was diagnosed with chronic lung obstruction. I'm so sick now that I have to lie in bed most of the time. Only once in a great while do I have the energy to go outside and toss a football with my ten-year-old son. My seven-year-old daughter draws pictures for me to cheer me up when I'm lying in bed all day. I used to love to read, particularly history, but now I can't remember what I just read. I have to keep going back to read a page over again. It takes me three months to read a little book.

One doctor recommended that I have a SPECT brain scan, which I did. The scan showed compression and diminished activity on the right side of my brain. Another test indicated that my blood clotting level is much higher than it should be. Tests for antinuclear antibodies have come back positive four times. My family doctor says that means I am likely to get an autoimmune disease like lupus. I already have two other autoimmune conditions—Hashimoto's thyroiditis and Sjögren's syndrome. Doctors have told me I have the body of a 60- or 70-year-old man, and I'm only 33.

I repeatedly went to see the VA doctors about a lump in my throat. For over a year I complained to them about this lump, but they ignored it, trying to say I was just having psychological problems. Finally when the lump grew, I was diagnosed with thyroid cancer and I had to have half of my thyroid removed.

After all these years, my private doctor and some of the VA doctors are saying I have Gulf War syndrome. They have referred me to the Gulf War syndrome clinic, but even though I received the anthrax vaccine and took the PB pills, the clinic has refused to see me because I was never deployed to the Persian Gulf.

The VA doctors keep saying Gulf War syndrome is stress related, but as I testified to Congress, I didn't even go to the Gulf War and my jobs weren't stressful. But I'm still experiencing the same symptoms as the other sick vets. I think the vaccinations and the PB pills made many of us sick. The French troops weren't given the PB pills, and not many of them are sick. So much for stress as the cause of Gulf War syndrome.

Now I'm once again having problems with a lump in my throat, so the cancer has probably spread to the rest of my thyroid. I wonder now if I will live many more years, and my main concern is worrying about my family. If I die from Gulf War syndrome and the VA hasn't recognized the condition as being service connected, then my family will get only a one-time death benefit of $800.

I gave the government five years of my life, and in return they have destroyed my health and haven't taken responsibility for their actions. It doesn't seem fair.

Marie

Designer

My husband and I had been in excellent health prior to 1997, when we had a professional pesticide applicator treat our house on the West Coast because we had an ant problem. He was supposed to use one pesticide outside around the perimeter of the house, spray the baseboards inside with a different pesticide, and also use a pyrethroid fogger inside. I had told him to use the least toxic products possible. He said that the pesticide he used on the baseboards was so safe they used it in hospital rooms and that the fogger was so safe it was used in senior citizens homes.

When the pesticide applicator came to do the spraying, we noticed that his canister was leaking on the floor. He told us that we could come back to the house in three hours and open up the windows and turn on exhaust fans to air out the house for an hour. We returned to the house in four hours, opened it up, turned on the fans, and aired it out for two hours before we went in to do some cleaning. When we went back in, the house smelled so bad that we had to put cloths over our faces. I thought to myself that must be how hell smells. It turned out that the house smelled like that for almost two years.

The applicator had said to wash just the countertops, but since we have grandchildren who live in the area, we tried to wash everything they could put their hands on. One of my daughters happened to have moved back in with us a month earlier, so she was helping my husband and me. After we had been in the house thirty or forty minutes, I noticed that my legs felt funny. My arms also felt strange; they were shaky. My exposed skin was also burning. When I said, "Does anybody else feel weird?" my husband and daughter said they were having the same symptoms. So we decided there was something wrong and grabbed whatever we could and ran out the door. We got in the car and went to a motel and just dropped into bed. We were experiencing nausea and chest pain, and our eyes were burning and teary. Our tongues were so thick they felt too big for our mouths, and we had a metallic taste in our mouths. We were also all salivating a great deal.

My husband went over to the house on Saturday, the next day, because we hoped the pesticides had evaporated, but it was just as bad. In the meantime the poison control center had said we should wash the baseboards with baking soda and water. On Sunday we all three went back to the house to clean some more to try and get that awful smell out of there. My husband got down on his hands and knees to wash the baseboards with baking soda and water. I believe that is when he got his overdose of the pesticides. He got so sick so very quickly that we knew we had to get out of there.

Our daughter wasn't feeling well and had lain down on her bed upstairs for a nap. When I called for her, she wouldn't respond. I ran up the stairs to her bedroom and found her unconscious. We later found that the concentration of pesticides was particularly strong in her room, and she was lying on an egg-shell foam pillow that had undoubtedly absorbed a lot of pesticide. I was feeling pretty sick and shaky myself by that time. I shook her until she finally opened her eyes, but she couldn't focus her eyes and she closed them again. I somehow gathered enough strength to shake her again. This time her eyes opened and rolled back in her head, and then her head fell backwards. I thought she had died. I don't even remember how I got her out of bed, down a long hall, and down a staircase to the front of the house and outside to the car, where my husband was waiting.

The next thing I remember is sitting in the car in front of the motel feeling so horribly sick I thought I was dying. We decided to go to the hospital because we were all so ill. That's when I heard someone say the word *poison*. I had been dry-heaving for quite some time, and they finally gave me something for that. They checked our vitals and told us they had called the applicator to find out what he had done to our house. They had also called the poison control center. After three hours, they sent us home. We were still so very sick we could barely walk.

The county agriculture department is responsible for applicators operating in the county. They denied that the applicator who had treated our house had done anything wrong. They came to the house four days after the application in order to inspect the premises. Their inspection lasted five minutes, and they did not take samples. I think they knew what they would find if they took samples.

We had our own private testing done about two weeks after the pesticide application. Even though fans had been running to vent the fumes out of the house, the tests showed residues of the two main pesticides on our bed, bureaus, drapes, kitchen countertops, baseboards, and carpets. A glass-top living room table looked like the applicator had taken a spray bottle and sprayed the entire table. There was a video on the table that was practically glued to the glass. When we had the house tested again two months later, they still found pesticide residues in some places. Even a year later, pesticide residues still showed up in my dressing room and on the living room TV.

We hired a professional cleaning service to go into the house five days after the application. They thought the house still smelled terrible and decided not to take the job because they feared what might happen if the cleaning chemicals were mixed with the pesticide residues. I contacted several cleaning services all over the country during my two years of research, but no one wanted to touch the job. A toxicologist from the company that makes one of the pesticides involved told me it sounded like we had walked into a "massive overdose."

Every time we went near the house after that, even just to open the door to let someone else in, we would start hypersalivating and have to spit and spit. Our noses would also start running, and all our other symptoms would return.

My husband is a brilliant man, but he is now functioning at less than 50% of his capability. Fortunately, he owns his own company, so he can go into work when he feels up to it. He has several national and international patents and has been honored several times by the government for his inventions, which are used by the Navy. Now, however, I watch him while he tries to tell the difference between the microwave and the refrigerator. I try to let him do as much for himself as he possibly can, except for driving. Once he could have been dropped in the middle of the woods in Maine, where he grew up, and he would have found his way out without even a map. Now he gets disoriented to the point of losing his way even in the house. He used to be a speed reader and now is unable to read above a third or fourth grade level. When I try to summarize what these poisons have done to him, I describe it by saying, "They have destroyed an entire library." When I saw his SPECT scan, I had trouble controlling my emotions. The brain damage is very extensive, and my heart aches when I think how brilliant he was and how in a matter of an hour some careless "splash jockey," as applicators sometimes call themselves, could have caused this brain damage.

Nighttime is devastating for my husband and nerve-wracking for me. He has night terrors, terrible nightmares in which he will yell and argue with someone and his blankets fly all over the place. He never used to have headaches, but now he does.

My SPECT scan and my daughter's weren't as bad, although she has more brain damage than I have. I was at the top of my high school class, and I graduated from college with high honors. Although I used to be excellent at spelling, I now have to rely heavily on a dictionary because I can't remember how to spell a lot of words. I was also a math whiz and could do any math problem in my head. Now I can't add two single digits without using my fingers or a calculator.

I'm an oil painter, but now I can't paint anymore because I can't handle the oil paints and solvents and my hands and arms are very weak and painful. I was also a fashion designer. My favorite project was bridal gowns and all the bridal party attire and floral arrangements. Now I can barely use a pair of scissors, and I can't handle the smell of the chemicals used in finishing fabrics. The designing part of my brain is also gone. Decorating a house was a cinch before, but now I can't decide which shade of white I want.

My husband and I can see that our memories are practically nonexistent. Thoughts cross our minds so fast that we can't retain them. My husband loses everything he touches, especially bills. Sometimes I give him something to mail, and two weeks later he brings it back home. We never know what day it is or what month. We've made big mistakes in dates. We tried handling things like business papers, but we realized it was time to ask for help, so we turned to our son, who by luck lives a few miles from us. He has been a godsend.

Ever since the massive pesticide exposure two years ago, I can tell when I've been exposed to certain chemicals because my body kicks right in. My nose feels like someone had sprayed acid in it. Then a headache starts right between my eyes and spreads to my brain, and I feel like my brain will explode and burn up. Other symptoms that I usually get are nausea, shakiness, tearing of my eyes, a runny nose, diarrhea, aches and pains in my legs and arms, weakness in my limbs, and abdominal pain. I had some migraines in the past, but now their frequency is greatly increased.

I often get reactions from going to department stores or other stores. It's also difficult to go to the houses of friends. I cannot visit the woman who had been my best friend since 1959 because she smokes and her house is saturated with cigarette smoke. The last time I visited her my lungs felt like they were on fire, and I was so nauseated I could hardly breathe. I was sick for days after that exposure. I've tried to explain why I can't go visit, but I haven't heard from her in over two years. We had brought up our children together. That was a very difficult and sad time, realizing my best friend did not care enough to at least stop smoking in our presence, or even believe us.

The pesticide exposure has caused a total role reversal in our marriage because my husband was even more affected by the incident. I'm now the head of the household. I pay all the bills as soon as they come in; otherwise I forget about them totally. I've tried very hard to retrain myself to remember where things belong, but it really doesn't work. Sometimes it just feels like an exercise in futility. It just causes more confusion and frustration.

After we were sickened by the pesticides, we found a relatively safe house to rent for two years, until the owners recently returned. There was a problem with ants there, as in our original house, but I finally got them under control with natural methods. On days when we had a nice ocean breeze, we could leave the windows open until someone lit a barbecue, a fireplace fire, or painted their house.

Unfortunately, the owners returned during the summer of 1999 to reclaim their house, and once again we had to find temporary lodging in a residential hotel, which added to the stress of our situation. Our children decided that we should go into the loony bin for a while. They thought we were suicidal because I happened to remark one day that I wished I could go to sleep and never wake up. We were told we needed relief from pressure and that the hospital would provide rest, support, someone to reach out to when we felt stressed, and someone we could talk to who would listen and support us.

Of course the hospital stay didn't turn out that way—it was just another hell. While we were in the hospital, all of our symptoms got worse because of the exposure to perfumed substances. We had agreed to spend 48 hours there, and we checked ourselves out after that time had elapsed.

Thank heavens we are now back in the residential hotel, where we can at least control our exposure to fragrances and other chemical exposures a little better. We are having our house totally gutted and redone to remove any lingering traces of pesticide and hope that we will be able to return there sometime soon.

Connie

Physician

I've heard that many people never have any idea how they developed multiple chemical sensitivity, but for me it was very dramatic. I had been doing my usual job; it was a Friday, and we had a very heavy surgical load. I'm a pathologist, and I worked in a hospital lab where I analyzed tissues from operations. Formaldehyde didn't bother me. I had been literally pickled in formaldehyde for over twenty years. I had never had any breathing problems, so I thought that after all these years, I was pretty immune to everything. This day was no different. I had been working on tissues wearing a mask, gloves, and apron for probably a good two hours straight, and all of a sudden I got hoarse. One moment I was talking on the dictaphone tape, and then my voice just faded away to nothing and I whispered, "I'll take a break." That was the last thing I wanted to do because it was a Friday afternoon. I wanted to finish; I was holding everybody up.

It seemed to be getting a little harder to breathe, and I never had had anything like that happen before. I said, "I'll go walk outside," and I walked outside in front of our hospital. Right next to that area was the door to the emergency room. Now instead of getting easier to breathe, it was getting harder. It was harder to take a breath in, and I started to make a funny noise that I had never heard before. It was like a croupy sound.

An ambulance had pulled up and let a patient off. The ambulance driver said to me, "Oh, gee, what's the matter?" And I said, "Oh, I don't know." I thought I had a frog in my throat, something like that. But he had heard that sound before, and he said, "I think you should be the one in the emergency room. You're sicker than the person we just brought in." I said, "No, I'll be OK in a few minutes. I just need some air."

Well, after a few minutes, not only could I not talk, I couldn't breathe. So he led me to the emergency room, and I knew that when they didn't have you sit down and talk to the triage nurse and go through all the insurance, it was serious. They whisked me right into a room, and all of a sudden I was the one lying on the stretcher. (Now, I'm used to being the one tending to everybody else. I'm a mother, I'm a doctor, I'm used to giving the orders.) I couldn't breathe, and I hoped they would know what to do for me. And so they started the IV line, and they were giving me steroids. They were doing this and doing that and the oxygen felt good, but anytime they would take the oxygen mask away, I would still have trouble breathing. So they called respiratory therapy down and started giving me nebulizer treatments with

Ventalin. I had never had an inhaler, I didn't even know what to do with one. But when they gave me the nebulizer treatment, all of a sudden within five minutes I could breathe again. The scary thing was that afterwards the doctor in charge told me they thought they were going to have to trache me (make an incision in my throat and insert a breathing tube).

Then they started asking me questions about what I worked with, and I kept saying, "But I've been working with this stuff for years." We tried to be very scientific, so I said, "Well, I'm working with formaldehyde, and now because we're worried about AIDS we have this bleach solution. Maybe the formaldehyde and bleach got together. Maybe I got gassed with chlorine." So that was logical, and I said, "OK, we won't do that again and I'll never have trouble again." I walked out of there thinking that was the end of it for me. I had figured out the problem scientifically; I was never going to have trouble again.

When I went home for the weekend, I told my husband, "Oh, a funny thing happened today," but I still wasn't taking the episode seriously. I had no trouble the whole weekend, but when I went back to work on Monday, odd little things started happening. I would be working with a slide and microscope and dictating, and all of a sudden I would get hoarse. So after a while my colleagues said, "It must be the carpeting in your room. It must have formaldehyde in it. Maybe we should get rid of the carpet." The engineering department said, "Oh, we'll clean the vents in the ceiling over your desk." And I don't want to tell you what came out of the vents when they decided to clean them, but I still had the problem.

So being a scientific person, I said, "We'll have to experiment." I started taking my slides to different rooms where I thought the ventilation would vary, and I still would get the problem. And then I said, "What do all these rooms have in common?" And it was the slides. So then I asked, "Well, what's on the slides?" A few years earlier we had switched solvents to something we had been told was very safe. Instead of xylene, which could cause cancer and also might affect the young women working in our processing labs who were of childbearing age, we said, "We'll go to this safer solvent." It was called limonene, and it had this wonderful orange smell that I loved. I had even heard that at some of the national conferences when they were first marketing this material, they had punch bowls and the reps were giving people glasses of it to drink and drinking it themselves and saying, "This is so harmless." I have now learned that no solvent is harmless.

So how did we figure out that it was limonene? Remember, I was working at an area where I was exposed to formaldehyde and bleach. Well, with a little more detective work, we found out that the people on the floor below us had been getting sick for almost two years—headaches, sinus trouble, burning throat, runny eyes—and they put up with it until a young woman got pregnant. When she told her doctor, he said, "You can't be around that." She literally searched the whole hospital until she found out where this smell, this very distinctive orange smell, was coming from. She found it in our lab in the back room where they would discard it all down the sink. However, the solvent was not soluble in water, so it would hang up in all the traps. As a result, on the floor below, the smell would come out their toilet and their sink. It was a small little bathroom, and they would get symptoms. So in order to keep the pregnant woman safe, our maintenance people said, "Don't ever throw the stuff down that sink in the back room." But without telling us, someone decided to throw it down the sink right next to where we worked. It turned out that on that fateful Friday, the week's supply had been dumped down that sink near me. Since I was working with formaldehyde and bleach, however, each of which had a strong odor, and I had a mask on, I never smelled the limonene.

The other interesting point is that the first time I reacted to the limonene, I had been working near the sink where it had been poured down for two hours or more before I had trouble. But later on the time interval between exposure and reaction kept getting shorter and shorter and shorter. It got to the point where I would lose my voice after two minutes of exposure. So being in a hospital, I went up to a pulmonary doctor, and he said, "Well, try an inhaler." I went over to the ENT doctor, and he said, "Oh, before you dictate your slides, try a little nasal chromolyn. You're probably having a little allergic rhinitis, so just squirt some of this stuff up your nose and you'll be OK." I have since found out that all that does is delay the reaction. And then of course, if you've already used your inhaler and you get a reaction, you have to wait for a certain time interval before using the inhaler again, so you've lost the protective effect of your medication.

Since I work in a hospital, after we realized what I was reacting to, we decided not to use that chemical any more. Once we stopped using it, everything was fine at work again, and I went for a couple of years without having any trouble at all. So again, I thought it was just something that I was allergic to, but if I avoided that substance, that would be it, no problem for life. Remember, I was young, I had my career, I felt invulnerable.

So what changed everything? Again, it was a very dramatic day. I had made a presentation at a lunchtime tumor conference and had come back to the lab, and now I was running late. My slides were ready and waiting for me (not with limonene anymore but with xylene as the solvent). But I was an hour behind the clock, so I was eager to start work. Now it turned out that in the storeroom they had found an old gallon jug of limonene, and rather than send it away to the company that took care of waste disposal for us, they decided to pour it down the sink to save $30. (We're a cost-conscious place.) They poured it down the sink in the histology area, which is probably five or six rooms away. It's an old building, however, and the vent systems are all interconnected. All of a sudden while I was working, my voice began to disappear, the croupy sound (laryngeal stridor) started, and I had a terrible reaction again. And I was still thinking: "Oh no, I've got to keep dictating. I'm behind the eight-ball." Then one of the other doctors came in and said, "I'm sorry, you're going to the emergency room." I said, "Oh, if I take a couple of puffs of the inhaler, I'll be OK." But that day because I think it was a whole gallon of limonene that had gone down in such a strong concentration, I really wasn't OK.

When I got to the emergency room, for the first time that I had ever been aware of, my blood pressure was sky high. My pulse was fast, and again, they were treating me in one of these serious modes. And it was after that episode that I was told: "You have to think of this now as if you had an anaphylactic bee sting reaction. You've got to get an epinephrine pen, and you have to carry it with you at all times." I still thought, however, that just a little inhaler would be enough, as it had been in the past.

As it turned out, there was a doctor in the emergency room who also had a practice outside—he only covered part-time—and he said: "I want to tell you that I'm scared for you. My son has bee sting reactions, and I think this is the time for you to go to somebody and really get worked up. If you don't know anybody, I know an allergist, and I'll send you." Well, I knew every doctor in my hospital. These were the people who had been my friends for years, but the minute there was the thought that it might be a workers' comp case, all of a sudden, my colleagues said things like: "Oh, I'm sorry, I don't take cases like that." So I actually had to go to the allergist that the other doctor had recommended. He did tests all over my arm and said, "OK, we're going to get to the root of this."

So I went through all the allergy workups, I did everything that he told me to do, and I thought I would be OK. I got through the next month, but something was different now. I would go to the tumor conference, and there was a nurse who wore a perfume that all of a sudden was giving me problems. Now before, if somebody had a strong perfume, even one I liked, I would get a headache, but now I started getting laryngeal stridor and I would get it very fast. (Stridor is the high-pitched whistling sound someone makes when they breathe when there is an obstruction in their larynx, trachea, or bronchial tubes.)

One of the things that bothered me the most when I started having all these reactions was the way it began to limit my activities. I like to do three things at a time and jump through hoops. I was a mother, I would run the kids around to all their sporting events, I was a good wife, I was a good doctor, and now all of a sudden, I was being wiped out. I was tired. In the past I always had a lot of energy, so for me to be tired now indicated that something was the matter. There was one day when I was dictating in the lab, and all of a sudden it was like word salad coming out. I couldn't get the dictation done, and fortunately another doctor I worked with helped me do it. That afternoon I literally could not even think of the diagnoses or put together a complete sentence. And that was another thing that was very frightening because it's one thing to have something affect your voice. But if you think something may be happening to your mind, that is very frightening.

One of my most serious reactions came on a day that was supposed to be my last day at work for two weeks because I was scheduled for knee surgery. I wasn't going to stay out any longer than I had to. I prided myself on getting to work every day and being there no matter what I felt like. So I was trying to finish up all my work before I left, and I was right down to the wire. I had one more case and I was focusing my attention on the slide under the microscope, so I didn't give any thought to what might be happening outside in the hall or across the way. As it turned out, the cleaning crew were there and they were doing the floor and cleaning the bathroom. Now in our particular hospital, the cleaning products have a nice orange smell. They were using a citronella-type product to mop the floor that is in the exact same family as the solvent I have trouble with, so I had a reaction. But I had one more case to finish up, so I took a couple of puffs of my inhaler and said, "Oh, that's OK, I'm going to finish the last case before I go." That was probably the biggest mistake I made. I've now learned that once something has triggered a reaction, you have to get away. You just can't stay. For me, avoidance is key.

So I finished the case and started walking out, but I was almost stumbling. I would say that was probably the worst that I had been. It was getting harder and harder for me to take a deep breath, and the croupy sound was getting louder and louder. You could hear it down the whole hall. I tried to leave and get to the stairway, thinking that I would be OK if I got some good air. By then, my head was woozy, my legs wouldn't work right, and I just couldn't take a deep breath. Fortunately, a nurse was coming down the stairs and got me to the emergency room.

That night in the emergency room something was different. I was there for three hours. We just couldn't get the cycle to break. I always thought that as soon as they gave you a puff of something, you got out of it and everything would be fine. But this reaction was lasting three hours, and my blood pressure went up to 220 over 110. I knew that was bad; you can have a stroke if it stays at that level. My pulse was racing, and I couldn't break out of the cycle.

After that third time in the emergency room, it was like one, two, three, now you've struck out. It finally hit me that this was something serious and I could possibly die. I went home, pulled out my will, wrote a whole addendum as to what to do with my children, my this, my that. I really felt vulnerable. That time I really thought I was going to die. And I still could barely talk, the whole night I could barely talk. I called the allergist up, and he realized something was different and he was great. He called me at midnight and six o'clock in the morning, and by the sound of my voice, he decided whether I should take more steroids.

The person who got up in the morning was not the same person who had gotten up in that house the day before. I now felt different in my own home. I couldn't tolerate the smells in my own house. When the dishwasher was running, the sweet-smelling steam was making me worse. When I tried to do a load of clothes, my own detergent bothered me. I tried to dry the clothes, and the fabric softener bothered me. Now none of these products had ever bothered me before, so this was the first clue that I might now have problems with something other than just a solvent at work. From there on in, literally nothing has been the same in my life.

Fortunately, I live in a state where there is a world-famous medical school. The chairman of occupational medicine had written about 100 cases of chemical sensitivity, so by June 1997 I knew what I had. We had a name for it. And of course I tried to look it up to learn more about it, but none of my textbooks included anything about chemical sensitivity. The people I worked with didn't know anything about chemical sensitivity either.

At the medical school, I was given the methacholine challenge test for occupational asthma, which was negative. I also had no reaction to the phenol preservative used in the test. However, something else in the room was irritating me. It was narrowed down to either the bleach solution or the antibacterial orange-colored cleaning solution. These were removed from the room during the methacholine testing. Afterwards they were tested individually, and I reacted with stridor to the cleaning solution, with the breathing pattern being documented in graphic form on the testing machine.

My local allergist could find no other case reports of limonene causing stridor, so I was the guinea pig in a double-blind experiment on two separate occasions. The histology supervisor made up vials, full strength and diluted, of the four chemicals I worked with most in the lab—formaldehyde, bleach, xylene, and limonene. The allergist had me inhale the samples through my mouth while my nose was clipped shut so that I couldn't tell what I was testing. The inhaled chemical that evoked stridor was always limonene in both full strength and diluted forms. The allergist graphed my reaction by spirometry. My case was written up as a posterboard display case report and was presented at a national allergy meeting. None of my allergist's peers had seen a case like mine caused by the solvent limonene.

It is worth noting that my pulmonary doctor asked what had altered my immune system around the time of my first attack. I remembered that ten days before that attack, I had received a flu shot. In fact, my left arm was still sore at the time of that first frightening attack that landed me in the emergency room.

To try to get a better handle on the diagnosis of my condition, I did go to a major teaching center and had a brain SPECT scan performed. It showed mild but diffuse brain hypoperfusion, which meant that I was not getting enough blood flow to the brain. It's scary to think that you're not getting enough blood supply to your brain, and that explains for me some of the cognitive and neurological symptoms that I get now. I'm probably most concerned about my brain. When I can't breathe, I get concerned about that, but it's more common for me now to have neurocognitive problems on a daily basis. If I avoid things, I can pretty much keep the breathing under control, but I can't change the neurocognitive. That has meant that after 30 years of working as a pathologist, I can no longer pursue that career.

To date, the cheapest and most effective therapy seems to be avoidance of bothersome perfumes, cleaning products, and other chemical exposures. This has resulted in major lifestyle changes for me. For example, mall shopping is now a rare event for me because my system can't handle the overload of perfume scents, new clothing with formaldehyde fabric finishes, and all the other chemicals one encounters in a mall. I also have to be careful what public functions or social events I attend because of the chemical exposures I will encounter in these places. My life has changed dramatically in many ways, but I have a strong Christian faith and believe that there is personal and spiritual growth and opportunity in even the worst adversity.

Richard

Painting Contractor

I was a painting contractor. I did that for about seven years. In retrospect, I have to say that I started having problems after about three years in that occupation, but I didn't recognize it until things really reached a critical point. Basically, I was having muscle spasms, and I would be driving down the road and I wouldn't remember where I was going or why. There were some other things going on with me at that time. My arms and legs would often feel numb. I also found that I was having a lot of difficulty breathing. I would feel a really heavy pressure in my chest; it was almost like having something inside that was pushing against the chest wall and it created a lot of pain. I also had a lot of pain in my spine, and altogether, I found it was harder and harder to breathe. I actually had to consciously force myself to breathe just to stay alert. A number of things like that were going on, and I wasn't quite sure what to make of it, and the problems kept getting worse and worse. Then a friend suggested that maybe I was allergic, so I went to see an allergist, and he recognized that I was reacting to chemicals.

After I started developing these kinds of problems, when I would go to do a painting bid and I would be trying to impress the customer and give him all kinds of information, that sort of thing, I would sometimes lose my ability to speak, and this was very distressing to me. I later came to associate that kind of reaction with exposure to certain perfumes and fabric softeners, but at the time I didn't know what was going on. All I knew was that I would lose my ability to concentrate well enough, lose my ability to think clearly enough to express myself. I would just stand there looking kind of stupid. Obviously I stopped winning contracts at that point. That went along with some other problems that I was having at the time. With those same kinds of exposures, my face would turn excessively red, beet red; that still happens occasionally. If I go in the grocery store and have an exposure to perfume, sometimes that will happen or I can turn red if they're using an ammonia floor cleaner. And then people stare at me, and I feel very uncomfortable.

Developing multiple chemical sensitivity has been the worst experience in my life because before that happened, I felt very competent at running a business, employing people. I felt in control of my life. I felt good about myself; I had a certain amount of authority, I had a number of employees that looked up to me. I just felt good about being able to pay my bills and taking care of my child support and everything else that's part of working and being a functioning member of society. One of the hardest things for me was to have my body essentially let me down on a continual basis over and over again and worse and worse, to the point where I just had to withdraw more and more to protect myself and keep myself from being excruciatingly ill.

I think especially as a man, not being able to work was very difficult. For all of my adult life, I guess I had identified with what I was doing for a job and had identified with my role as a business owner, someone who was providing employment for other people, someone who was performing a service for people, someone who got a big fat paycheck when the job was done. I had kind of a tradition—I would always take myself out to a steak dinner every time I finished a job. Well, eventually, I had so many employees and so many jobs going on at one time that I was going to be having a steak dinner every night, so eventually I stopped that ritual. But I felt really good about the whole process. In my last couple of years as a painting contractor, however, things started going downhill. I was earning less and less, and I was having more trouble thinking clearly enough to organize things and keep things on a good track. It was hard for me to see that slipping away, but it was a lot harder once I began reacting to everything in a very overt way, a way that was very devastating.

I used to be very involved in social organizations and community activities and that sort of thing, but that's something I've not been able to do for about the last seven years. I'm now trying to take a computer class. The first time I tried to take a computer class five years ago, it didn't work out very well, but it's working out better now. I still have problems that I have to deal with, mostly involving exposure to cosmetics. Every time I come home from class, I have to spend the rest of the day just trying to recover from the exposures there, but the teacher was very nice about not wearing cologne or aftershave after I talked to him about my condition. I didn't want to make a big deal out of it, so I didn't talk to the other students in the class. Occasionally someone will have on an aftershave or a perfume, and by the end of the class, my face will be kind of red. I have a hard time speaking after I've been in the room longer than an hour, and that's partly because my voice gets really weak and strained. I also can't organize my thoughts well enough to ask a question after I've been there a while. What I usually do is write down my questions and bring them in and ask the teacher at the beginning of the next class. I have to find ways like that of dealing with problems I encounter in the process of trying to accomplish something I want to accomplish. I want to learn how to use a computer so I can connect with people, try to find some way to get connected with this world.

As it is now, I feel somewhat isolated from other people because it's hard for me to interact with others face-to-face because of the scented products they use.

One of the things I find especially difficult is that I have to work ten times harder now than I used to just to accomplish ten percent of what I used to get done (and used to take for granted that anybody could do). And it's frustrating that other people don't understand. They just see the ten percent that I'm able to accomplish and they make all sorts of assumptions and judgments about why I'm not doing the other ninety percent. It's frustrating that the condition that I'm dealing with is so unknown, and there's so much misperception or misinformation about it. That's unfortunate because I think there is a need for education even among people who are only maybe five percent along the way toward developing a really serious chronic condition. They could save themselves a heck of a lot of heartache down the road by avoiding the kinds of chemicals, the kinds of exposures that are going to lead them down that same road I've been on. There's just such a huge gap in understanding chemical sensitivity in society. I know that it was hard for me to understand it when I first started reacting in a profound way to environmental stimuli. That didn't fit in with anything I had ever learned. It didn't fit in with my way of experiencing the world and perceiving my life, so it was very confusing for many months until I really got a grip on it.

One of the things that's has been most helpful to me in trying to get my nervous system back onto an even keel is avoiding the exposures to solvents, to perfumes, things like that which trigger the problems in a major way for me. In fact, it's the only thing that I've found that allows me to be relatively comfortable. As long as I avoid those exposures, I can be relatively comfortable at home, but as soon as I go back out into the world, if I go to the store, if I talk to somebody I run into on the street, chances are I'm going to end up feeling pretty sick afterwards. It may be because of something people have washed their clothing in or perfumes they are wearing. I'm very sensitive to perfumes. When I come home from the store after running into someone wearing perfume, I'm sick for several hours. I really would like to get back to the place I was before any of this happened, where I could just go to the store and my thoughts wouldn't have to be focused on avoiding someone who is walking down the aisle with perfume on. My thoughts could be centered on what I was doing in my business or a presentation I was going to make at a social group or something like that. It's very frustrating to be in the position of not being able to do anything, not being able to go anywhere basically without suffering tremendous consequences.

Tina

Teacher

In the early spring of 1996, I was teaching biology at a high school that was undergoing major renovations that were scheduled to occur over a three-year period while school was in session. During the early spring months, demolition started taking place. The entire gymnasium was being leveled, and a pile-driving operation was occurring outside my window. All that spring, I had severe coughing episodes and difficulty breathing. I had never been an ill person or been one to run to doctors, so I treated the condition myself. In retrospect, I can see those problems were the beginning of the end.

During the summer months, however, my health greatly improved, and I was able to work as a naturalist on a whale watch boat for the entire season. Then it was September, and the first day of school arrived. Within minutes of entering the building, I had an excruciating headache and difficulty concentrating. Several teachers voiced their concerns about the foul smells in the building, but we were reassured that there was nothing wrong.

Within a week, I was being treated by my doctor for severe migraines. I was told to take the medication he prescribed for me once a headache started and then every hour until it diminished. On a typical school day, I was averaging five or six pills, with no relief. I returned to the doctor within a week to voice my concerns and say that I thought the fumes from the school renovation were triggering my headaches. He indicated no interest in the possibility, probably because of the political ramifications if citizens thought something was wrong with the school. At that point I found a new physician.

During September and October, I noticed that I was feeling great on the weekends but would develop symptoms sometime Monday morning after I returned to the school. My symptoms included headaches, burning eyes and nose, an inability to concentrate, and dizziness. When I went home each day, I was so exhausted I had to go right to bed. I would sleep for six or seven hours, get up for a little while and eat something, if I wasn't feeling too nauseated, and then go back to bed. By mid-October, I noticed that when I used a window cleaner on the weekends, I would develop some of the same symptoms that I had at the school. And by the end of October, I was losing my voice within twenty minutes of the time I entered the building. This of course made it very difficult for me to teach since I could only whisper.

During this entire time, construction was proceeding. Gym floors of a rubberized material that contained isocyanates were poured while we were in session, and demolition was carried out by men wearing ventilator masks while we taught in the adjacent rooms. The onset of allergy cases in children skyrocketed. High school students who hadn't had allergic problems before were now wheezing in class. Several of my students were out for weeks on end with symptoms like mine, as well as bloody noses and unidentifiable rashes. Some students who used to get good grades were now making C's. Several students had to use inhalers to enter the building, but when the administration was questioned, the reply was, "There is nothing wrong" or "It's all in your head." In November, I had the union attorney meet with interested staff to discuss environmental illness and their rights. Although several faculty members had complained of similar symptoms, very few people attended the meeting because they were afraid of losing their jobs.

On December 12, the superintendent came to talk with me about some school matter, and I was unable to speak to him because my voice had disappeared as usual when I entered the building. He decided at that point that I should no longer be teaching and sent me home. (My department had already repeatedly told him about my voice problem, to no avail.) I was 28 and the first teacher to leave because of health problems related to the constant renovation. Two days later, another staff member, who taught two doors down from me, was sent home by her specialist.

Finally, some of the students held a walkout to protest the unhealthy conditions at the school. The administration closed school early for Christmas and performed several air quality tests (which they claimed they had been doing all along). Unacceptable levels of isocyanates were found and became the scapegoat. Even to this day, some physicians in town will not acknowledge a person's problem to be school related if isocyanates are not known to cause that particular symptom.

School reopened in January after a major cleanup, but three other teachers left by March. I was put on workers' comp in December 1996 and remained on it until September 1999. By the time I left the building, I had become sensitized to many common chemicals. I had to start using nontoxic cleaning products in my home and buying unscented personal care products. My husband works on ocean liners carrying petroleum products, so he had to take off his outer clothing and boots and seal them in a plastic bag before he entered our apartment. Then he had to shower immediately. I had to do my Christmas shopping out of catalogs, and I didn't buy a pair of shoes in three years. The exposures in stores were brutal. My school-related symptoms would appear almost immediately when I entered a store. A bad exposure in a store would bring on the same extreme fatigue that would make me sleep all afternoon and all night. Meanwhile, I was being treated by a specialist dealing with occupational and environmental illness. He was extremely thorough but also admitted that there was no medical treatment and told me my condition was not accepted by most physicians.

Realizing that I needed to do something to reclaim my life and educate the public, I started talking to anyone I could about environmental illness. The paper ran a few articles following my life while I remained out of school. The sick building situation at the school was an important issue locally because not only were teachers sick, about 20 students could no longer attend the high school because the building made them ill. Tutors had to help them with their classes in a building at an elementary school. Several students and teachers now have MCS as a result of their exposures to the ongoing renovation, and many others have developed respiratory problems like asthma.

By staying out of school for a couple of years, doing some helpful alternative therapies, and eating a strict and healthy diet, I have improved greatly. To my surprise, I even got pregnant, although I had been told that it was unlikely I would ever be able to conceive. We have a very normal and healthy two year old. Am I totally better? By no means. I can enjoy dinner with my husband in a restaurant for the first time in two and a half years, but I will never be able to enter a smoky pub setting (that's not the end of the world). I can't drive with my car windows down in traffic (maybe I could on open roads, but not around here). I can enter most stores now, but I know my limits. I have permanent, irreversible damage to a vocal cord, so I have to wear a wireless mike when I teach. Another teacher has to use a wireless mike now, and at least five other staff members complain that they lose their voice in certain rooms or areas of the school.

I am now teaching full-time again for the first time in three years. That makes me very happy because I love my occupation. I am a member of the health and safety committee at my school that evolved from this horrible ordeal. This week I finally got the school department to replace all the dry-erase markers with water-based ones that don't contain harmful solvents. I am fortunate that the current school administration allows me a lot of freedom that most don't enjoy. Some advice for anyone in a similar situation: (1) Keep a journal of your symptoms, including the places they occur, dates, and times. I did that from the start and it has been invaluable. (2) Put everything in writing and always keep copies. (3) Keep asking questions.

I would not wish on my worst enemy the chemical sensitivity problems I developed working in my high school, which clearly qualifies as a "sick building," but I have learned from the experience and have grown as a person. Fortunately, my health has improved enough that I am not afraid to dream anymore.

Editor's note: Effects on the vocal cords resulting from chemical exposures are not uncommon in MCS. These effects may be transitory, resulting in a temporary laryngitis, or they may involve more serious damage to the vocal cords. An article in *Teacher Magazine*, October 1998, described the health problems occurring in a California school bordered by strawberry fields on three sides. Spraying of pesticides was commonplace, and the spray often drifted onto the school grounds when students were outside on the athletic fields. One teacher who was caught outside in a cloud of spray got a headache and a burning throat almost immediately. When his eyes swelled shut, he went to the emergency room, along with several other teachers. Since that exposure, he has had eight operations on his throat. (He was a nonsmoker.) One other teacher reacted to the spraying episodes with a numb mouth, headache, and sore throat. She had to have a cyst removed from her throat. Two teachers in the school have had their larynxes removed because of cancer. At least one of these two teachers was a nonsmoker.

Michael, who contributed the first story in this book, has discovered that if he uses a certain gun cleaner, his voice suddenly becomes very high pitched and squeaky.

Gunnar Heuser, M.D., Ph.D., F.A.C.P., is particularly interested in vocal cord damage and covers it in his protocol of suggested tests that should be performed to document chemical injury.

Ariel

Graphic Artist

In 1989 I was diagnosed with MCS. My doctor believed that it had started when I was exposed to an improperly vented computer printer that used liquid toner. For three years I had been unknowingly trying to cope with what is now known as sick building syndrome. The extreme fatigue, the disorientation, the eye-focusing problems, could all be traced back to the printer exposure. Each year the symptoms had grown progressively worse, developing into a pattern of increasing health problems with each exposure to gasoline, newspapers, exhaust fumes, cleaning supplies, or particleboard.

My husband and I decided to take early retirement, and we moved from Denver to a picture-postcard village in western Colorado, completely unaware that I could be affected by the pesticides that were used on the orchard across our back fence. Each time the farmer sprayed his trees, I reacted with symptoms that were frighteningly severe—face and hand numbness, a total inability to focus my eyes, an extreme pressure in my head that made me feel like my face would blow off, chills that seemed to turn the marrow of my bones to ice and produced uncontrollable shivering, and sudden feelings of depression and despair. These initial symptoms would gradually give way to a chronic flu-like state that involved bone-wracking aches, extreme fatigue, and difficulty breathing.

We soon learned that I could avoid these extreme reactions by leaving the area. We asked the farmer to notify us prior to each spraying to give us time to get out of Dodge City, as my husband was fond of saying. But since the farmer was spraying three or more times a week, it was hard for him to keep us notified. At any rate, by taking advantage of the wind patterns, we soon learned that we could spend mornings at the house, long enough to shower and get clean clothes and put together food for our evening meal. Then we would spend evenings and nights in the mountains. Even with all our attempts to stay out of the house when the wind was blowing from the orchard toward our property, it was hard to avoid all contact with the pesticides.

Because the severity of the reactions I suffered with each spray event indicated that future exposures could result in anaphylactic shock, my doctor recommended that we move to an area free of pesticides. So in 1991 we moved to a beautiful valley where we felt safe behind the twenty miles we knew separated us from the nearest agricultural areas, where there might be herbicide or pesticide spraying.

We bought a house and made the necessary changes to create a nontoxic home, adding a large artist studio for my work. We planted extensive gardens of vegetables, fruits, and berries to assure a pesticide-free food source. I was content to spend my hours working in my garden and my new studio, hiking the hills around our home, enjoying the animals and all the beauty that nature had to offer right out our back door.

We thought we had put a reasonable distance between us and the pesticides. But I wonder if I will ever again make the mistake of thinking there is a line behind which I am safe. In 1997 my illusion of safety fell apart when an extensive noxious weed containment program was instituted in our area. It was clearly impossible for me to stay at my home because weather patterns could keep pollutants circulating in the valley for weeks. Hence I joined the ranks of the dispossessed and homeless. With an overwhelming sense of gratitude for my nontoxic camper and with a respirator always close at hand, I roamed the back roads of national forest areas for ten long weeks. And with an anaphylactic reaction to pesticides always a possibility, the shadow of death followed me every step of the way. No longer can you set up camp with the assurance that the area has not been sprayed, no longer can you settle in with a good book with the knowledge that a spray truck won't appear on your doorstep. There is no place to hide. At best, we merely bought some time.

We had invested considerable time and money in creating a nontoxic home in which I could once again be a functioning member of society. This home was my sanctuary; I cannot stay in motels, cannot even stay in homes of relatives or friends. Herbicide use in the valley would force us to sell our home and move. But where would we go? Where would we even begin to look? We couldn't just bop out and buy another home because it is almost impossible to find an area free of pesticides or other air pollution, a home free of toxic building materials and insecticides.

In the search for a new home, I came to know full well an overwhelming feeling of desperation, and along with that desperation came the growing conviction that the chemically sensitive are viewed as "throw away" people. After many miles and dollars and weeks of searching, in May 1998 we at last found a house we thought I might be able to tolerate. We purchased it, but only after assurances by the owners that they had never allowed pesticides on the place. The house needed minor alterations to remove particleboard, and in the process, access to the crawl space was created for a few plumbing renovations. There in the crawl space we found numerous cans of ant insecticide bombs. By calling the EPA, we were able to determine by the lot number that the previous owners were indeed the ones who had placed the cans under the house. Because there was no way to decontaminate the building to make it safe for me to live in, we were faced with the decision whether to try and sell the place and again attempt the impossible—finding another property that could possibly meet my needs for a pesticide-free environment—or to hang on and build a new house elsewhere on the property.

We finally decided to stay and build a nontoxic house on the property. Winter caught us, however, without a habitable house. We were forced to shelter in our safe Airstream trailer and camper. Neither were built for winter habitation, however; the water systems would not withstand freezing weather. That winter we had to get our water from the main house a hundred feet away, and we used the bathroom and laundry facilities there. I would have to go in and out quickly, however, because the house was too contaminated by pesticides for me to spend much time there. But even though we had no running water or bathroom facilities in the trailer or camper, I was very thankful that we had them. I could remember all too well the days before we moved here when we were escaping from the pesticide spraying in a neighboring orchard and spent too many days and nights with no shelter at all. At least the camper and trailer provided warmth and shelter from the wind and rain. And most of all I am grateful for a loving husband who takes on my troubles as his own and perseveres to a solution.

The necessity to build a new house has required us to cash in the last of our IRA's and other savings accounts. But aside from the monetary problems, it has been extremely hard to live with the fact that our lives can be disrupted to this extent by a noxious weed control program. It's taken a large amount of mental fortitude and energy to pick up and start all over, to trust that we'll get it right this time and will not have to move again. We have once again cleared the brush, put up a deer fence, tilled and planted a large garden, planted fruit trees and berries, and we are building a new house. Once again we are facing the dilemma that so many MCS people and their families face: spending down their income and savings to accommodate their illness, with no guarantees, no protective laws, no safety nets, and little basis for hope.

Timothy

Ten-Year-Old Child

In 1988 my husband and I adopted a healthy baby boy from Calcutta, India, and we named him Timothy. From the time we adopted Timothy when he was four months old until he was four years old, he was very happy and healthy. He had only one ear infection during that period.

Then on September 18, 1992, we had wall-to-wall carpet installed in our family room. Two weeks later we noticed that Timothy was getting sick; he was getting a chronic cough. His pediatrician couldn't find exactly what was wrong, but he put him on codeine. During this period Timothy was spending a lot of time in the family room lying on the new carpet and coloring. After a while, his breathing became very bad. He couldn't walk, and he couldn't talk. At the age of two, Timothy had been able to spell both his first and last names, but now, when he was four, he couldn't even say his name. Then one night Timothy developed severe vomiting. It wasn't regular vomit, he was vomiting mucus and it was choking him. I held him all night, rocking him, and I called his doctor four times for advice.

Within a few days Timothy had become so sick that I had to carry him upstairs and to the bathroom. He didn't have the strength to walk, and his arms hurt and his joints hurt. He would just curl up in a fetal position and lie on the couch. When I told this to Timothy's doctor, he told me that Timothy was playing mind games with me and that he wasn't sick. He said Timothy just didn't want to walk to the bathroom. He told me he thought Timothy was lazy and that he should get up and get moving. And this is a four year old we're talking about.

We then took Timothy to a specialist, who said he had severe asthma but he didn't know how it had started. Timothy was put on ten different medications, and he was stable but he wasn't getting any better. In the meantime, he was still playing in the family room, lying on the carpet. Then we saw a television show about the problems new carpets can cause, and we took him to a pediatrician who specializes in environmental medicine.

We think that Timothy reacted to the carpet more than my husband and I did because he was spending so much time playing in the family room. He would lie on the carpet near a radiator that was blowing fibers and fumes from the carpet right into his lungs.

Of course we decided to have the carpet removed. Two of the men who did the work became very ill while they were working and had to leave. One of them ended up in the hospital for three days with breathing problems.

We had Anderson labs test our carpet, using four mice.[1] Two of the four mice died. The other two mice showed gross behavior; they were trying to eat each other and were walking backwards. We also had a well-known chemist analyze the carpet, so that we were able to find out what chemicals were in it. The chemist said there was a very high level of formaldehyde in the carpet, and that is a chemical that can cause severe problems.

Since Timothy has been diagnosed with severe asthma, we have to be very cautious about what chemicals he comes in contact with. Exposure to formaldehyde, perfume, cleaning products, or wood smoke will often give him an asthma attack. He is still on medication, and there is no indication that he will ever outgrow the asthma. Fortunately, his mental functioning returned to normal after we got rid of the carpet.

We were told by a specialist that because Timothy's motor skills had been impaired by the exposure to the chemicals in the carpet, we should try to get him into some special activity to improve these skills. My husband and I checked into karate and found an instructor who was willing to work with Timothy every day initially so that he could learn to do jumping-jacks and the basic hopscotch routine. The karate has helped Timothy recover his motor skills, and he is now very good at karate and wins trophies.

I find it very hard to go into the store where we bought the new carpet that produced severe asthma in Timothy. I'm concerned about the other parents that I see looking at wall-to-wall carpeting because I worry that they are going to put it in their family room or nursery.

---

[1] Anderson Laboratories, Inc., of West Hartford, Vermont, performs experiments in which they expose mice to air blown across new carpet and then check for changes in the breathing, behavior, ability to move, and other physical characteristics of the mice. They do similar experiments using air fresheners and samples of air from a school. Their website is
<<www.andersonlaboratories.com>>

Donna

College Professor

In 1957 my mother was accepted to the college where I now teach, but she was unable to attend because she could not afford the tuition. In 1995 we were all excited when I was hired as a professor at this small liberal arts college. I was 27 years young, my husband Paul was 29, and we had found the perfect old farmhouse to rent. The farmhouse was far back from the road and bordered a wild game reserve. The deer came right up into our yard, often in groups of three or four. I used to watch them out of the kitchen window.

The day that my husband and I moved into the house we smelled a gas odor in the kitchen. When I called the gas company to report the leak, I was relieved to hear that there was no need to worry—it was just a residue of the "chemical that they add to the gas to make it smell." When the tanks are low, I was reassured, you can smell the odor but it does not indicate a leak. When the gas company filled the tanks, the smell got stronger, but I was told that filling the tanks had stirred up the chemical that they add to the gas to make it smell. The man from the gas company installed an oven safety valve in the stove to calm my fears.

During my first winter living in the farmhouse, my toes would go numb for periods of time, often when I was doing the dishes. My vision seemed to be getting worse, and at times I found it hard to read my books or grade papers. Sometimes my hearing seemed bad, and when I talked on the telephone, which was in the kitchen, it seemed like people on the other end were not holding the receiver close enough to their mouth.

It was in 1996, eleven months after we began unknowingly living with a gas leak, that low levels of chemicals in everyday products began to trigger my symptoms. My sinuses often began to swell not only when I was in the house but also in highly polluted areas such as the New Jersey Turnpike. In July 1996, I fell down while I was dancing but had no memory of falling and had not had an ounce to drink. As my husband put it, "You were just there and then you were gone." When it is hot and humid now, I can't be around perfume without experiencing sudden blackout episodes—I'm there and then I'm gone.

By the fall of 1996, I was experiencing numbness in my toes, vision disturbances, hearing difficulty, and incontinence when I sneezed. In addition to these symptoms I was having regularly in the house, at school I was often feeling confused while I was teaching. I couldn't recall words or names that I was extremely familiar with, and I began to pull up similar sounding wrong words frequently. I began to think of myself as Mrs. Malaprop or Archie Bunker. I had also begun to consider myself somewhat of a klutz because I was always tripping over things and bumping into corners of doors instead of walking around them. And I was only 28 years old.

During my parents' Christmas party in 1996, I was talking with my new step-mother-in-law and I could not remember basic words like the word *car*. I could think of what I was trying to say, but I could not find any words to say it. The more I was around this woman, the worse it got. By the spring of 1997, when I was near her I could no longer even think of anything at all to say; it was like someone had taken an eraser to my brain. My husband and I joked that it was the ghost of Millie, my deceased mother-in-law. Without joking, I wondered if I felt so ambivalent about having a new mother-in-law that it prevented me from being able to speak. The temptation to resort to simple psychological explanations for physiological problems induced by chemical injury is great indeed. It turns out that when my step-mother-in-law does not wear her favorite brand of designer perfume, I have plenty of things to say. Clearly not ambivalent feelings but chemicals.

In the late spring of 1997, I also began to experience a sharp pain in my lower abdomen, and my vision disturbances took a new twist: I would see a zigzag at times in my upper right visual field. It turns out that this is a chemical-induced "migraine without the migraine." I get the migraine vision disturbances but no headache.

On August 1, 1997, my sister and her three children moved into the house with my husband and me. We decided to put them all in the upstairs bedrooms, and my husband and I began to sleep in the bedroom next to the kitchen, where we could still smell gas. I began to experience a persistent cough, shortness of breath, chronic fatigue, and muscle weakness.

By the time October arrived and we began to close up the house for the winter, my shortness of breath had become so extreme that I couldn't even finish sentences. On October 15, 1997, a sharp pain in my lower abdomen awoke me from my sleep, and I felt the urge to pace about anxiously. As the lower abdominal pain went away, my abdomen swelled so much that I looked as if I were eight months pregnant. From that time until we moved out of the house in late December, I had all the symptoms described above, plus I began to experience ringing in my ears, shooting pains down my legs, nausea, vomiting, tingling sensations, and palpitations. I would get odd phrases stuck in my head, and I began to have a crooked look because my right eye appeared to bulge. I remember looking in the mirror and thinking that I looked like the self-portraits done by insane painters.

As the severity of my symptoms increased, I began regular visits to doctors' offices, emergency rooms, and urgent care facilities. But they were looking for disease, and I was experiencing poisoning. They were never going to understand what was making me sick.

Over the two and a half years that we lived in the house, I called the gas company many times to report my concern about the odor. I had them check for a leak several times, but they always said it was fine and safe. I did not know that they had never conducted an air quality test to check for the leak. Finally, they told me that the odor came from a residue in the tanks because the tanks were old, so we had the tanks replaced. We could still smell the odor, however, and the gas company still made excuses.

In desperation, we finally moved out. When we moved into our new house on December 23, 1997, the dozens of symptoms I had been experiencing had begun to subside to a handful. Then on January 5, my sister and I went back to the old house to do some cleaning. When we walked in, we commented to each other on the gas odor. Moments later I sneezed and experienced incontinence. My abdomen swelled again as we cleaned the house, and I felt awful. I was anxious and belligerent.

As it turned out, a few days later our former landlord smelled gas in the kitchen of that house. When he asked the gas delivery man about the odor, the delivery man pulled the stove away from the wall, and there they could see a crack in the flexible copper tubing, the crack that the gas company had assured us for over two years was not there.

I at once called poison control in New Jersey and learned that the first symptom listed for propane exposure was numbness. Immediately I thought back to when the numbness in my right toes began just months after we moved into the house. The next symptom was tingling. Boy, had I tingled. Out of the thirteen or so symptoms listed, the only ones that I had not experienced were cyanosis, coma, and death. When I called the gas company to let them know that they had missed the leak for two and a half years and the exposure had caused me serious health problems, I only wanted to prevent the same thing from happening to someone else. The gas company representative could not have cared less. No big deal, he told me. There are no ill health effects from breathing leaking gas for over two years—none.

Even though I did not believe his assertion, I was actually relieved to finally know the source of my many symptoms. I was feeling better and better and had no idea that I might have developed chemical sensitivity; I didn't even know it existed. I figured that it was just a matter of time before my blurred vision and tingling sensations would also go away.

Then in late January 1998, I went to teach a class in the art building, and many of my old symptoms returned. I figured that there must be something in the building that was like propane, so I had my classes moved to another building. I thought, "No big deal, I just can't be in the art building." My thinking was fuzzy in the afternoons again, but I tried to ignore the lack of attention and forgotten words. I was still expecting my health to keep getting better and better, but then chemical exposures began to bother me. When my sister put bleach in the washing machine, I suddenly felt short of breath, had lower abdominal pain, and had abdominal swelling. When I tried to go into a dry cleaning shop, I could not breathe. When the students began to use formaldehyde in the science building, I felt as though my throat was closing, my speech became slurred, and I again had abdominal swelling. Once again I had to move my classes to another building.

In early March 1998, I flew to Minneapolis for a conference, and I had a severe reaction during the flight. I had to be given oxygen when I arrived. I could not focus or concentrate at the conference. The chlorine smell from the hotel pool made my abdomen swell up and made me short of breath. The perfume and aftershave the other guests were wearing made my speech more impaired, and the carbonless-copy-paper receipt the front-desk clerk gave me literally took my breath away. When I got back home, my abdomen was still swollen and I felt very tired and sick, so I called the doctor I had begun seeing in early February. We agreed that it sounded like I was experiencing chemical sensitivity. He said he would find an expert in the area of chemical injury, and through a series of references I ended up with a physician who understands the condition. I called for an appointment with the naive belief that she would have some kind of detoxification cure. While she is great, I learned quickly that there is no way to reverse this condition totally. The best way to remain functional is through avoidance of the triggers that make the symptoms worse.

I have been more fortunate than many others with chemical injuries in that my family has been very supportive and so has my employer. We have replaced our previous personal care and cleaning products with nontoxic, unscented alternatives. We have replaced the gas stove, and even though we freeze, we leave off the gas heat. My husband took a crash course in installing some hardwood floors that come with a water-based finish. As the air quality in our home improved, my health improved. The accommodations that my family members have made have allowed me to remain a part of my family.

Fortunately, I have been able to obtain accommodations at work through the Americans with Disabilities Act. My employer made a special office and classroom for me with ceramic tile floors, windows that open, filters, and all metal furniture. My students are required to be free of perfumes, colognes, scented skin lotions, fabric softeners, and freshly dry cleaned clothing. My entire floor is cleaned with unscented, nontoxic cleaners, and the grounds-keeping crew must notify me of any pesticide applications on campus.

Yet even with all of these accommodations, I am only able to teach in the mornings, and every day that I work on campus I have urinary frequency and blurred vision. Often scented products used by other professors cause me to have incontinence, memory difficulties, nausea, vomiting, or diarrhea. I have not presented a conference paper since 1996, nor have I published since then. An editor agreed to publish my dissertation as a book with some revisions, but my attention span is so impaired that comprehending new theoretical material is impossible. While I have been lucky in obtaining accommodations, I fear that luck will not carry me through the tenure process. My chemical injury has made tenure very unlikely, and if I don't get tenure, I will become an unemployable Ph.D. in my early thirties.

I am most fortunate in another way, however, because I now have a beautiful, healthy baby girl. Suzy is six months old and was conceived five months after I moved out of the house with the gas leak. I asked for and received accommodations at the hospital for the labor and delivery. I had to labor and deliver in a specially prepared room with unscented staff, an enormous filter, a ceramic oxygen mask with tygon tubing, glass bottles for the IVs, specially rinsed linens and baby blankets, and all surfaces rinsed of the residues from cleaning products.

So here I am at 31 years old, lucky to have a wonderful baby, husband, and supportive family. I'm fortunate to have accommodations at home, accommodations at work, accommodations at the hospital, and accommodations even at my own baby shower. Yet I have white matter in my brain (qMRI), lack of blood flow to parts of my brain (SPECT scan), and severe attentional deficits (neuropsychological exam), among other things, and the only known effective treatment to stop further deterioration is avoidance of the triggers. We all know, however, that avoidance leads to an isolated life. So I go back and forth between retreating from a world that now makes me sick nearly all of the time and working for ADA accommodations that allow me to occupy a small part of the world that I used to roam about in freely.

Has it really only been four years since 1995? What a long, strange trip it's been.

Tomasita

Housekeeper

Back in February of 1993 I worked for my employer as a housekeeper, and I went into her guest house on a daily basis. On February 18, I went into the guest house to get a magazine for her. I smelled a faint odor of perfume, but I didn't think anything of it because she had had some guests staying there. My brother Moises had gotten sick the day before while he was working there, but we just thought at first that he had suddenly gotten the flu.

Later that afternoon when I went back to the guest house to mop the kitchen floor, I could still smell perfume. I got my bucket, filled it with water, put in some floor cleaner, and started to mop the floor. While I was working, I turned on the trash compactor, and then the smell became more intense. I continued mopping the floor, but I noticed that a lot of muscle tension was setting in. The right side of my face became numb, and I had a feeling of being very fearful. I finished mopping the floor and went back to the main house. When I told my employer what had happened, she told me I could leave for the day. When I was walking out the door, I started crying and became anxious and excited.

I got in my car and started driving home, but I became very disoriented. I knew I was going home, but I didn't know if I was going the right way. I started salivating a lot, salivating like a bull dog. Finally I made it home, and I took a shower, which made me feel a little bit better.

The next day I went back to work, but I didn't go back into the guest house. I still wasn't feeling well. I felt very irritable and had a lot of muscle tension. The following day was a Friday. I had decided to go back into the guest house to empty the trash compactor. When I picked up the bag, I started having tremendous muscle tension in my neck. The right side of my face became numb, as if I had had dental work done. I became fearful again, and the salivation started again. I went back to the main house, and my employer told me to go home.

That weekend we went out to dinner because it was my birthday. I got very sick in the restaurant, so we had to leave. The rest of the weekend I was just out of it. I didn't have no energy, pizazz to do anything.

The following Monday I went back to work. My employer had decided that I should not go back into the guest house, but I started reacting when I was near other people who came back from the guest house. I felt like something was crawling all over my skin. Then as the weeks went by, I started feeling that way in other places. I started having difficulty with my speech; I started stuttering a lot. Certain places I would go I would become very fearful, even angry, become violent. I knew something was wrong with me, but I didn't know what. I had gone to urgent care, and they didn't know what was wrong. Finally my employer arranged for me to go see an allergist, and he told me that my problems came from exposure to a toxin.

As time passed, I was even reacting to things in my own home. I would sleep outdoors because I felt better outside. I was at my wits' end. When people came to visit, I would get sick. (I didn't realize yet that the chemicals in their cosmetics or clothing were bothering me.) When I got a reaction, my gums would swell, my tongue would swell, I would have difficulty concentrating, I would itch a lot, and salivate.

When I started having reactions in my home as well as at my employer's place, I thought that I might have brought home some sort of bacterial or mold contamination from her guest house. So I cleaned my house with industrial strength disinfectants. I washed the ceiling beams and the walls and the floors with disinfectant. Before long I couldn't tolerate being in my house anymore, and I couldn't use any of the blankets from my house. Now I realize that it was the detergents and disinfectants and soaps that I was using that were continuing to make me sick.

Finally, my doctor recommended that I see a specialist in environmental medicine, so at last in April, I consulted a doctor who understood what was going on with me. By now I was no longer able to go into my house without getting sick or to go in anyone's house.

Since I felt better outdoors, I tried living in a tent at the construction site for my brother's house for about a month. Then I had to leave because I was sensitive to the construction materials stored on the site, and I also became sensitive to the tent and couldn't sleep in it anymore.

Finally my brother Moises got a cleaning crew to come to my house, and they cleaned everything with plain water. After we had aired out the house for a few weeks, I was finally able to come back into it and live there again.   I had to learn what cleaning products and personal care products I could use that wouldn't make me sick.

My daughter was young, she was about ten years old when I got sick, and she thought I had gone crazy. They all thought I had gone crazy because they weren't bothered by the things that I was reacting to. Eventually when I finally got to a doctor who understood my situation and I started reading literature about chemical sensitivity and we began changing our lifestyle, people in my family became more knowledgeable and understanding. My ex-husband wasn't that understanding, however, so he left me. The hardest part of being sick was finding someone who understood the situation and could say, "Yes, there is something wrong with you. You're not crazy." Finding that doctor enabled me to get back to a point where I can live basically a normal life, except that I have to be very careful of exposures.

There are still many things that I'm very sensitive to. I make a point of going to the grocery store early in the morning or late in the evening so there aren't so many people around, with their soaps and perfumes and personal hygiene products that bother me. Even when I go out in the car, I avoid the main streets and take all the back roads because car fumes bother me a lot. If family members want to be around me, they have to use things that are perfume free. Nobody is allowed to smoke in my house.

I have to be careful where I go. There are a lot of peoples' houses that I can't go into, or if I go into their houses, I can't stay very long. Some people understand, and some people don't. Some people think it's all in my head, but my family and I know it's not, so that's the important thing.

———————

Over the years since that exposure in the guest house, we've thrown around so many different theories, wondering what made me sick. I've always thought that maybe a can of pesticide was thrown in the compactor because when I turned on the compactor the smell became more intense. Lots of pesticides have perfume added to cover up the pesticide odor, so that would explain the perfume odor that I kept smelling. I've always speculated that maybe the guests that were there sprayed some pesticide around and threw the half-empty can in the compactor. I've always felt something was put in the compactor, but the bag is gone, so we will never know.[1]

———————

I had a long battle with workers' comp, and when we finally came to the hearing part of it, the judge basically told me it was all in my head and I was just trying to get rich and he was not going to help me get rich. It was very frustrating. My daughter broke out in tears and told him he didn't know what he was talking about and when it happened to someone in his family, then he would. I had to calm her down, but we were both in tears because it's very frustrating to have somebody tell you it's all in your head when it's not.

———

[1] Excessive salivation, which was one of Tomasita's most prominent initial symptoms, is a hallmark symptom of pesticide poisoning.

Moises

Caretaker

When I was growing up, I had never heard of multiple chemical sensitivity, but as a teenager, I did have problems tolerating strong perfumes. Mentholatum bothered me a lot—my grandmother often used that—and I just did not tolerate rubbing alcohol. Lots of times we used gasoline as a solvent to clean parts, and when I did that I would get a headache or develop a rash and itch. It just didn't set very well with me, so I tried to stay away from gasoline.

In 1993 I was employed as a caretaker. My employer had two condominiums; one was a guest house for family or friends when they came to visit. One day I went over to the guest house to pull the shower curtains so the housekeepers could come in and start laundering the curtains and cleaning up. I went into a bathroom off the master bedroom and reached up and started undoing the shower curtain. Suddenly I had the feeling of being gassed, and I felt like something was encompassing me. I got very shaky, almost frantic. My head was spinning, and I had this feeling of anxiety.

I went back next door to tell my employer. By that time I was burning from head to toe and was nauseated. I was salivating and just frantic, almost hysterical, not knowing what was going on. My employer had me sit down and gave me some water, but the salivation kept up and I became very irritable. I had never reacted to anything like that, and I didn't know what I was exposed to. And along with the other symptoms, I felt like I had just taken a chug of perfume or something like that because my mouth tasted like perfume. I was salivating, I was nauseated, and my heart was racing, but that was probably the most awful part of it—I just tasted the perfume in my mouth. It was so intense it was frightening.

My employer dismissed me, and I went home. I live out of town, and I had to stop twice on the way home because I was sick to my stomach. I was disoriented and turned into the first entrance to the area where I live, not the second one, which was my usual route. I was ready to crawl out of my skin and still had the taste of perfume in my mouth. I was irritable, ready to kill. I just could have slapped anybody that would have got in my way for anything. The salivation, nausea, the perfume taste were just horrible.

The next day I felt like I'd been hit by a train. My body just ached everywhere, and I couldn't move. My brain was still in a cloud, wanting to work but not really catching the gears and doing what it was supposed to. My short-term memory was just lost. For the next few days I was just a zombie. I would shower frequently because of the burning-type itch on my skin. That was the only thing that would soothe it and give me some relief.

Then a depression set in. I had such a horrible depression for months to come after that. We had company from out of town to visit, and I was just not social at all. I stayed in my room all the time they were here. The rest of the family entertained them and took them out and around. It was just a real bad depression; I was frightened, paranoid. I've never experienced depression like I had after the exposure at the condominium, since or before or now.

In the late sixties I was in Vietnam as an infantryman in the Marine Corps and was in combat quite a bit. We used tear gas launchers, big packs we carried on our backs that had numerous little rockets in them. They generated tremendous heat when we fired them. But the experience I had with the exposure in the condo that day affected me more than being in that type of situation in Vietnam.

At least I rebounded from the exposure in the guest house better than my sister Tomasita did. Prior to her exposure, she wore lipstick and perfumes and used cleaning products. Now she's just way more affected long-term than I have been. She's much better, but she's still has some very, very difficult times. Cigarette smoke bothers her tremendously, and she can't tolerate certain buildings, while I can to a certain degree. I'll go into a building and know something is getting me, but I'll do my business and get out. With Tomasita, a small exposure like that sets her back.

For years Tomasita worked as an EMT for the emergency medical services. She was around all kinds of people, chemicals, hospitals, clinics, that sort of thing. She also did domestic housework for many people here in town for years. She was able to handle cleaning products, cats, dogs, and peoples' smoke and that sort of thing. But after this exposure in the condo, she just can't tolerate those things anymore.

Throughout my experience with this exposure, I've had a great education, a forced education. I wish traditional medicine would recognize chemical sensitivity in some way and not turn their backs on you, saying it's psychological, that it can't happen, because I've seen it happen in this city and state. I've seen it happen to people from all over the country—children, adults, old people, poor people, rich people—and they're all in the same boat. I just wish that traditional medicine would take a closer look at it and start accepting it as being there. It's happening. There's a lot of people becoming sensitive or are and don't recognize it yet.

Randa

Planning Staff Analyst

I worked for a land-use planning agency in San Francisco for ten years, and across the hall from my office was a state cosmetology lab where beauticians came and took their state board exams. They used acrylic nail chemicals and applied acrylic nails daily, so I was exposed to low levels of acrylic nail chemicals and other cosmetology chemicals over a period of ten years.

There was also constant renovation going on in the building where I worked. Staff would frequently become ill, nauseous, dizzy, and get headaches. When we would go on our lunch break, we would feel fine, but when we came back into the office, we would start having symptoms again. These remodeling projects went on constantly for the ten years that I worked for that agency.

Then I came to work on a Monday morning to find that they had glued down carpeting over the weekend, and my life changed that day and has never been the same since. I walked into the office and felt as if somebody had taken a baseball bat and hit me across the head. I immediately started having very severe reactions to the carpet. I couldn't think straight, I couldn't talk straight, I couldn't walk straight, I couldn't compose a simple letter. I had pain in my chest, swollen lymph nodes, and burning in my lungs. By the end of that day, I felt like I needed to take some time off, stay out of the office while the carpeting was airing out, so I took an emergency vacation leave of three weeks.

I packed my car and drove to Oregon to visit my parents. I noticed that during the long drive there my breathing was very labored. I was also very fatigued. I got to my parents' place and had been there a day or two when my mother came down from a shower with some scented body lotion on, and again I felt like someone had taken a two-by-four and just hit me across the head. The body lotion smelled like she had poured a whole bottle of perfume all over herself, and of course she hadn't. Later she opened a bottle of nail polish in the bathroom while I was in there, and I had that reaction again. The same thing happened when she was cleaning the sink with some chlorinated cleansing power, and I thought, "What is going on with my body?" I was terrified.

When I came back home and went back to work, the reactions just continued to build and increase. I came to work, but on weekends I could hardly get out of bed because I was so sick and severely fatigued from all the exposures at work. I would sleep all weekend so I could go to work the following Monday. I could barely climb the stairs to my office. I was so exhausted that I had to pull myself up by the hand rail. I started reacting to copier fumes in the office, to the smell of the copier paper, to the shampoo people had used, to their laundry products, to the files that I had to use, which smelled moldy. We had a new secretary who wore a lot of perfume, and I could barely walk by her without feeling like I was going to pass out. Finally, I got an air filter and put it in my office, and I would hardly ever leave my office. I had people bring me things, but that was a problem because they would come into my office with scented laundry products or perfume on their clothing. I would react to those things, and I would have to call them on the phone and say, "Gee, could you please not come in my office anymore? Something on you that's scented is making me sick." In the subway on the way home, I could now smell the grease on the subway tracks. I suddenly felt as if I was becoming another species. I had this increased sense of smell, about 20 times the normal human sense of smell. I was smelling all these things, and not just smelling them, I was having bizarre reactions to them.

By that point I was reacting to the gas heat and the gas stove in my house, so I turned off the gas. I bought electric space heaters, an electric toaster oven, and a hot plate. I ripped out the carpeting, which was bothering me because it was moldy. I was starting to react to mold, which had never bothered me before. I had never had an allergy in my entire life to anything at all—molds, pollen, foods, nothing would make me ill. I had no idea what it meant to have any problems with anything. I was very healthy, I had gone hiking in the Himalayas. Then I became a person who could barely get out of bed because I was so fatigued. I had to wear an industrial respirator to go into the grocery store because the fragrances on people in the store would make me so ill I couldn't figure out what I was doing there, what I was supposed to buy.

I kept trying to work, but my doctor said, "You must take a leave of absence and get out of that environment to see if you can regain your health." So finally after about six weeks of continuing to work after the carpeting was installed, I took a six-month leave of absence.

At the end of this leave of absence, I tried to go back to work, but immediately when I walked back into the office, I knew it was an impossibility. I couldn't tolerate even walking down the hallway into the office. Cleaning products were a problem, as well as many other things and people in the office. My sensitivity was so heightened that it was impossible for me to be in that work environment at all, even with the air filter and the respirator. I just couldn't do it. My boss was complaining that I couldn't compose a simple letter that made any sense. I went back to my doctor, and he said, "You really can't work there anymore." But that was my career, I had worked there for ten years. When I had to leave, I didn't even get to say good-bye to anyone, to the people that I had known for ten years. Management was very unsympathetic; they were pretty cold about what I was going through.

I applied for workers' compensation. They denied my request and said, "The illness you claim to have isn't recognized and therefore we don't recognize it, and therefore you're not sick and you're not entitled to anything." And I thought to myself, "Boy, this is pretty crystal clear to me. It's like one day I'm fine and the next day I'm totally ill and disabled." It took me nine years of litigation to get workers' comp.

Over the years I and my fellow workers in that office had frequently noticed solvent vapors coming through the air vents when remodeling was going on somewhere in the building. Management was quite belligerent when we complained and just ignored us for the most part. Finally we said, "Well, why do they insist on doing this work during the regular work hours? They should be doing it on the weekend or at night so we're not exposed to these fumes." And management said: "Oh, it would cost too much money if we did the work in the evenings and on weekends." The reality is, it would cost whom too much money? It would cost them too much money, but in the end my exposure to toxic chemicals in my workplace has cost me my life, it has cost me my health. I haven't worked for over nine years now, and that really does a job on your self-esteem.

Living with chemical sensitivity is being chronically ill and feeling crummy most of the time. After nine years of that, it really wears on you. It's really tiring, and I don't know how to explain this condition to people. It's drudgery and monotonous and lonely and isolating, and your old friends and your family don't want to hear about it, and I don't want to hear about it, but it's my life. I would like to take a vacation from this as well. It's just very painful, and there's not a whole lot of joy left in my life. This is my life, however, and I try to live it as best I can. I get pleasure from listening to good music. My greatest joy in the world has been nature and the outdoors and hiking, but since I got sick, I am now so sensitive to pollen and mold that it's rare that I can enjoy hiking. I also don't have the stamina and the energy that I used to have for those kinds of things, and I can't go camping now.

Sometimes I dream that I'm healthy, but I wake up and say, "No, I've got this bizarre illness." Living with it year after year, sometimes it just seems like I can't take it.

Randa's update, October 1999: I have been living in Santa Fe, New Mexico, for six years, and I have been looking for environmentally safer housing for myself for a good part of those six years. Those of us with MCS look at dozens of houses, call about hundreds more, and still can't find anything that works. The problems in finding "safer" houses that work for people with MCS are fresh paint, new carpet, heavy fragrances, pesticides, mold, gas heat, close neighbors who use pesticides, too much interior raw pine wood, toxic cleaning products or cigarette smoke residues, newer houses that emit formaldehyde and other toxic construction chemicals, or houses that are too old and are moldy.

My particular problem in finding housing here in Santa Fe is that the majority of houses are pueblo-style architecture and have flat roofs that typically leak when the houses are over ten years old. The leaks cause irreparable mold in the roofs. If I walk into a house that is over ten years old, it's almost always too moldy for me and my lungs tighten up immediately. I've looked at dozens of houses over ten years old, and mold has been a real problem for me in all of them.

I really haven't thought seriously about building my own house because many people with MCS who have built their own houses have had disasters and have not been able to move into their places for a long time (sometimes years). I am so exquisitely sensitive to building materials that it would be very hard for me to build a house that I could move into immediately. At least with an existing house, I can walk in and get an idea whether it will work. I did recently visit Snowflake, Arizona, and thought the houses that people with MCS had built there seemed quite good, but building a house is still something I hesitate to do. The financial risk is just too great if it doesn't work.

I presently live in an apartment that I own. I had no desire to buy an apartment because of potential problems with neighbors, but I couldn't find anything else that seemed like it might work for me, and I had to move out of a house I was renting a few years ago. I was fortunate enough to have some money from my workers' comp lawsuit and from my parents to buy my place. My apartment has hot-water baseboard heat, an electric stove and hot water heater, and tile floors. There are no fragrances and no pesticides in the apartment. Since it is a downstairs unit, I don't get any mold from the leaky roof. The apartment has worked fairly well for me except for the fumes from the exterior laundry room and occasional exposures to neighbors' incense, cigarette smoke, or barbecue lighter fluid.

Unfortunately, my upstairs neighbor installed new carpet at the beginning of summer, and that made me homeless for over three months until the carpet had outgassed enough that it was tolerable for me. As soon as the carpet was installed, my apartment filled with toxic chemicals that made me very ill and I had to leave right away.

I had to sleep on friends' porches for many weeks. The whole experience was extremely stressful, and I had no idea if I would ever be able to move back into my apartment. Fortunately, I was able to buy a used van a year ago with some of my workers' comp money. After extensive work replacing the interior vinyl upholstery and padding with several layers of cotton fabric, I was able to tolerate the van enough to sleep in it. After I got tired of sleeping on friends' porches, I decided to try taking a trip to California in the van. As it turned out, I had quite a nice time. I slept in my van for six weeks, often in neighborhoods near the beach. I hung out by the ocean and cooked my food on a camp stove in a beach park. It was great. I was able to visit old friends I hadn't been able to see in years because their homes are too moldy for me to stay with them. I won't say it was easy to travel this way, but it was worth it.

I am so much more fortunate than many people disabled with MCS. I have enough money to live on, and my family is going to help me buy a house if I ever find one I think I can tolerate. Many people I know who have MCS live on under $600 a month and live outside in tents or in their cars year after year. One friend nearly died from hypothermia. I am grateful for many things that I have in my life, even if it is a hard life. I appreciate every single kindness from people; that helps balance out the occasional person who reacts with disbelief about chemical sensitivity.

Liz

Registered Nurse

Until three years ago, I was full of life and enjoying it to the utmost as I worked at my job as a nurse in the distant reaches of northwestern Canada. Then suddenly I was hurled into a nightmare from which I have yet to return.

On October 1, 1996, during a night shift at my workplace of ten years, a staff member cleaned the bathroom that was located three or four feet from me, as she had done many times. This time, however, the door happened to be open. Being a super-conscientious cleaner, she poured bleach into the toilet bowl, used a foam disinfectant-cleaner on the toilet seat, ammonia glass cleaner on the mirror, and another strong cleaner to clean the tub. At the time my attention was focused elsewhere when suddenly I felt strange and turning to her, asked, "What the hell are you doing in there?" The words weren't even out of my mouth yet when I felt like I couldn't get enough air into my lungs and felt like I was choking. I ran to open all the doors and windows to get enough air. I call this experience my "blast." It seems probable that the problem arose from the mixture of ammonia and bleach because it is well known that the combination produces a very toxic chlorine gas that can be lethal.

During the next three months, I became more and more sensitive to more and more things: exhaust, diesel fumes, cleaning products, marking pens, shampoo, deodorants—the list went on and on. At first things were problematic only at work, but before long I started having reactions to chemical exposures elsewhere. Although I am a nurse, I didn't know what was happening to me. I thought I was going crazy, except that I was usually able to track down the source of my difficulty and I became better once it was eliminated or I left the area. I was major scared.

During this period, we were directed to use bleach full-strength to clean the bathrooms, in spite of my difficulty with it. It was finally discontinued when, on the advice of my union, I threatened to refuse to work under unsafe conditions, citing the Occupational Health and Safety Act and our collective agreement. Within one day of the bleach being discontinued, my condition improved dramatically.

Shortly thereafter, however, I was working one day in mid-November when I noticed some sort of diesel or furnace fumes in the building. By the end of December, many other staff members and patients were complaining of headaches from the fumes of uncombusted fuel oil that were emanating from the air exchange vents all over the building. Looking back, I can see that I was the proverbial "canary in the mine." But when I detected the early warning signs in November, everyone just laughed, rolled their eyes, and said, "It's just that Liz again."

Upon investigation, maintenance personnel found that the exterior air intake vent had completely iced up from the winter cold, preventing fresh air from entering the building. To make matters worse, the exhaust fan in the basement had been turned off and covered with plastic to keep out the cold. We learned that this fan should be on at all times to provide the oxygen necessary for proper furnace fuel combustion.

Once the ventilation system problem had been cleared up, my health again improved dramatically and immediately. By this point, however, I had been assaulted by toxic substances too much and too long, and I couldn't sustain this temporary improvement in my health.

For some people, their sensitivities first become apparent when they start reacting to foods, but I first noticed reactions to chemicals and later became sensitive to many foods. By the end of December, I was able to tolerate only five fruits and vegetables. I was very sick and lost 20 lbs. in one month. My doctor was worried that I was becoming malnourished, and I feared I was headed for another world if things continued this way.

Before my "blast," I had had no allergies or sensitivities, save a slight intolerance to lactose and one medication. Three months later I was reacting to everything and was unable to maintain adequate nutrition. I now stayed away from people—friends, groups, everybody—as much as possible because everyone smelled of some fragrance from their shampoo, hand lotion, perfume, laundry detergents, or fabric softener. Friends did, and still do, find it hard to understand or believe how sensitive I am to fragrances, and most are unwilling to accommodate me by using unscented products. Previously I went out a lot. Now I hibernate in the sanctuary of my own home as much as possible, living the life of a hermit.

Even stores and restaurants bother me with their fragrances, cleaners, and new clothing odors, which send my body reeling. Sometimes I don't even have to be aware of the odor of some chemical for it to bother me. Frequently, I have only found out later what had been around at the time I had a reaction.

As my condition progressed, my symptoms included feeling weak, shaking, being unable to think straight or to put a sentence together properly, pressure in my sinuses, tightness in my chest, visual disturbances, welts on my tongue, and burning in my mouth.

Fortunately, when I attended a union health and safety conference quite early in my illness, I found out about the harmful effects of chemicals that can result from their improper use and storage, as was the case in my workplace. Eventually I was diagnosed with multiple chemical sensitivity, and I learned that though there is no known cure for MCS, the best way to control the illness is to avoid triggers that cause symptoms. I vigorously pursued avoidance because I was intent on licking this devil of a condition. While I have yet to get cured, life has become noticeably more manageable as I have cleared my house of all cleaning products and fragrances.

I also try to keep problem substances out of my workplace, but this has not happened without a fight. With the union's help, I have fended off persistent attempts by some of my co-workers and employers to sabotage my ability to continue working. They would intentionally sneak into my area and hide or use offending products on the sly to see if I would notice anything. Then they would follow me around mocking me and implying I was crazy as I tried to figure out what the problem was and remedy the situation. This exhausting, long-standing battle has improved considerably, but still persists to this day, three years later. Things may start to get better soon, however, because others in my workplace are also becoming chemically sensitive.

I am tremendously grateful for what I have today. I have made great strides since those first three months back in 1996. While my illness is still very disabling in many ways and probably always will be, I now feel well enough to venture forth into stores, some restaurants, meetings, conferences, and other venues. I do pick and choose carefully when and where I go. I now have a safe home with an air filter attached to the furnace, and this helps me to clear my reactions more quickly. While many people cannot seem to accept the validity of MCS, even when they themselves are sensitive to some chemicals, others, including some of my co-workers, are coming around and are now willing to accommodate me.

Although there are still many foods that I cannot eat, my dietary tolerances are now much greater than they were back in those days when I could eat only five fruits and vegetables. I also look healthy again. Many individuals with MCS cannot work, but through all this experience I have kept working. Granted, I might not have gotten as bad as I did had I been able to get a reprieve from my toxic workplace, but I was not a permanent employee, so not working was not an option. I still have my bad days, but now I experience more days when my reactions are quite tolerable and manageable. In fact, when my sensitivities are accommodated as much as possible, I don't even know I have a problem. While there are still a multitude of things I can't enjoy, there are more and more that I can. I am improving steadily and still refuse to give up the belief that one day I will be well again, no matter how long it takes.

Abner

Chemical Engineer

The following story is excerpted from testimony that Abner gave before the Governor's Committee on the Needs of the Handicapped in Santa Fe, New Mexico, in 1996.

I've had MCS since 1984. Since 1984, I've had twenty different employments. I've been looking very hard for steady employment, and it seems to be elusive. Two things are very difficult for people with MCS. One is that there are many jobs we just cannot do because of new carpet or carbonless carbon paper or new equipment or employees wearing perfume. Another problem for people with MCS is that when we have a job, sometimes we have to resign our job because of changes in the environment that take place.

I graduated as a chemical engineer in 1971. I worked for five years for two different state governments in pollution control, and then I started to work for 3M. I enjoyed working for 3M. Some of the technical problems were very interesting to work on, and the staff was very well informed about technical issues. When I acquired MCS in 1984, 3M tried to accommodate me. They transferred me from a chemical pilot plant to an office job, but at that particular time, I was too sensitive to function well and I received a very bad performance evaluation. In the same office another employee had been transferred from a laboratory to an office near me. Another gentleman who was a director and had had twenty people reporting to him now works by himself in an office that 3M constructed for him in which he controls air flow. He is fortunate to be still employed by 3M.

I had several jobs, and I went back to school and got a teaching certificate in Minnesota, and I have done substitute teaching there.

I've applied for many jobs. I applied for a job with the city of St. Paul which would require me to work with carbonless carbon paper. After the interview, I told them I couldn't do the job. Shortly after I left 3M, I applied for a job as a consulting engineer, but I couldn't wear the clothes, nor could I work in the building with new carpet.

Just last week I applied for a job in the Questa elementary schools, and the superintendent, Señor Gonzales, was very kind to me. I went to him before the interview, and I told him I had multiple chemical sensitivity. He said, "Let's go up and see the building." It was a job in Red River. He introduced me to Lee Gonzales, one of the principals in Questa, who took me into the building. Within a couple of minutes, I knew that I would get dizzy and disorganized, and I would be irritable with the children. I couldn't justify being a teacher and being irritable with the children because of the physical structure of the building. It's just not fair to kids. So I told Señor Gonzales I couldn't do it. He was very kind to me, and when we drove back, he gave me a history lesson on New Mexico which I very much appreciated.

In other cases, there have been jobs I had to quit because new carpet was installed. I had a job as a counselor in a halfway house and had to quit the job when carpeting was put in. I was a personal care attendant, and when the family remodeled the house and put in new carpet, I had to quit. I was flipping burgers and running a cash register, and I got a very good performance evaluation after two months and after four months, but when they hired new employees who wore perfume and kept scheduling me with them, I had to resign.

I do substitute teaching, and when I substitute teach, every day when I come home I have to wash my clothing to get the perfume off.

Thank you very much for coming. I appreciate your being here.

Editor's note: Six weeks after he gave this testimony, Abner took his own life. Lynn Lawson, editor of the newsletter of the Chicago MCS group, told me that two or three weeks before he ended his life, Abner called her to ask if she knew of any housing anywhere in the country that would be safe for MCS people.

Unfortunately, Abner's suicide is not an isolated case. By chance I happened to hear that within a three-week period, Abner and a woman with CFS/MCS on the West Coast took their own lives and another woman with MCS ended up in the hospital after a failed suicide attempt. Two people with MCS with whom I have happened to speak by phone during the last few years have committed suicide and one other has attempted suicide. When I asked in my survey of people with MCS if respondents had heard of MCS suicides, one man wrote this:

Yes. It is fundamentally disturbing to me to relate that a very good friend of mine, a dear friend, committed suicide some few years ago. She was young, maybe 30. She was exquisitely sensitive and finding a reliably safe place for her to live was almost impossible. Her biggest problem though: no money except the minimum Social Security income. Thoughts of her suicide still make my mind go numb. I myself will commit suicide sometime in the next few years. Why? Too maladaptive with no money as offset.

Time is running out for too many people with MCS. To convince them that it's worth holding on a few more months or years, it is important that society move toward an acceptance of the reality of MCS. This change in attitude would promote both badly needed medical research and attempts to help those with MCS find safe housing. Nontoxic housing would give those who are severely chemically sensitive a chance to regain their health and reduce their level of chemical sensitivity to the point that they could live more normal and productive lives. For further information about the need for MCS housing and ways in which contributions or investments could be used, please write:

MCS Information Exchange
2 Oakland Street
Brunswick, ME 04011

Bonnie

Graphic Artist

I was born and grew up in Cleveland, Ohio, at a time when it was a very polluted place. I remember that when we children swam in Lake Erie, my mother would wash us off with a liquid detergent when we got out to get the oil spill off of us. When we sailed on the Cuyahoga River, we would drop napkins and paper plates overboard and laugh as they dissolved from the acids in the water. So I had many, many chemical exposures as a child, but they didn't have an effect on my health until I was older.

When I was 17 years old, I was a dancer in a ballet company; at this point we lived in Minneapolis. I started having tremendous pain in my feet, and once we started looking into it, we realized that I had severely damaged the cartilage doing point dancing. After a couple of years of trying to fix it any way we could, a well-known clinic put artificial joints in my feet when I was 19 years old. The joints were a silicone product and have since been recalled because they were a failure and couldn't even bear the requisite weight. I was the youngest person ever to have these put in, and my body started rejecting them right away. I was a college student and could barely walk because my feet were swollen and red all the time. Within a couple of years, I went back to the clinic and said, "Please, this is a problem." They wouldn't go near it, so I went to my county hospital and had the joints removed, which is probably why I'm alive today. However, I believe that all my chemical sensitivity today stems from having that silicone in my body for several years.

I still had difficulty walking because at this point I had no joints in parts of my feet, but I went on with my life. I finished college, got a teaching degree, moved to New York City, and became an illustrator. My health held for a number of years, although I wasn't as strong as other people and I had more pain. I got sick more often than others, but I was still functioning. At one point I went back to a physician and let him do fusions on the bones in my feet because the pain was so severe. I'm sure that the anesthesia and the pins in my feet added to my exposure load also.

At this point, working in commercial art in New York City, I was exposed to many chemicals. I had my hands in benzene all the time, and I supervised the printing of my pieces in printing plants, so my load was getting higher and higher. The joint disease that affected my feet spread to my back as degenerative joint disease, and I had to have two disks removed.

After the birth of our first child in 1990, we bought and renovated a house in Maine. I knew nothing about safe materials, and we used many questionable products, including new carpet treated with stain-resistant chemicals. We also made the house very tight. It was my first year not working, and I spent the year lying on the floor playing with my daughter inside of a chemical-laden house. It took about a year for full-blown MCS to develop in me. It happened gradually, and it was kind of like a horror show watching it because I could not stop it.

I began developing various symptoms, muscle spasms, burning eyes, tightening of the chest, rapid heart beat, body numbness, gastrointestinal distress, hormonal changes, skin welts, brain fog, short-term memory loss, and adrenal response. The illness went from sensitivity to specific substances that were benzene-ring hydrocarbons like petroleum, benzene, formaldehyde, and silicone to generalized sensitivity to many common things like molds, foods, and fragrances. This was in 1992. We tried every treatment available. Most made me worse. At this point we decided to relocate to the Southwest to at least escape from the mold in Maine. Also at this point, I had my second child, who was born healthy but allergic. The pregnancy destroyed what reserves I had left. I could no longer take supplements or medications, and I could eat only a few foods.

The warm, dry weather of the Southwest has definitely helped in the course of my healing, but there are several things hampering my recovery at this point. First and foremost is the widespread use of pesticides in the Southwest. We have been forced to leave our house several times and sleep in the hills to escape exposures. We have had to move on one occasion to escape tremendous hostility from neighbors who resented our request that they not spray pesticides or at least notify us when they did so. A second problem is the lack of good medical care. I find there is still no understanding of this illness, which I believe is a mixture of endocrine, neurological, and immune system dysfunction. A third problem for me has been the recent failure of another joint in my feet. I am in tremendous pain, and as a chemically sensitive person, I do not and cannot take pain medication, and surgery is almost impossible because of the exposure to anesthesia.

In terms of a life with MCS, I find it extraordinarily difficult. The life limits are devastating. I can't work, there are very few creative outlets for me as an artist, and there are no cultural events I can attend unless they are held outdoors and at a park that doesn't spray. The social isolation is profound, and if you relocate for your health, there is a sense of distance from family and friends that is also profound. This illness is a terrible burden on the healthy spouse, who must make one life change after another in order to continue living with the sick spouse. It's also very difficult for children to grow up with a sick mother. They have awarenesses that they would never ordinarily have and fears that they shouldn't have.

On a good note, I've been working to help open a charter school in our community. We opened this fall, and it is the first public school in America to have a written nontoxic policy. They paint with safe paints, they do not use pesticides, and they use safe carpet cleaners. It's absolutely wonderful to be able to send my children to a school where I know they are safe.

Carina

Executive Secretary

Before my health problems began, I held down a good job as an executive secretary for a large railroad. I took wonderful vacations, had a very active social life, and made good friends. Then my health started to slowly deteriorate in 1975 after I moved to a brand new apartment that had particleboard cupboards, cement floors covered by new carpets, and a gas furnace in the bedroom closet. I stayed in this apartment for three years while my health continued to get worse.

In 1979, I bought a condo, and as soon as I moved into it, I felt terrible. Thinking that the former owners might have had pets, I had the unit exterminated. I couldn't stand the smell of the pesticide, which lasted some weeks, so I replaced the old carpets with new ones. I did a few more renovations in the unit and felt worse after each one. I didn't know what was happening to me and honestly thought I was losing my mind. I was unable to sleep more than two hours a night, my work at the office suffered, and going out socially completely drained me. Then my office was completely renovated, and I eventually collapsed. I was expecting to return to my job after taking a few months leave of absence to try and determine what was wrong with me. That was 17 years ago.

In 1981, I was diagnosed with environmental illness (or multiple chemical sensitivity, as it is usually now called). By that point I was reacting to many different chemicals and had to flee my pesticide-contaminated condo and leave all my clothing and furniture behind me. I moved into a stripped-down apartment with no furniture and slept on a pallet on the floor. I couldn't read because I was so sensitive to printers' ink, and I couldn't watch TV because of the fumes the wiring gave off.

During this period, I was living on savings and waiting and worrying every day whether my application for disability would be approved. Thank God, it was. I was able to find someone to rent my condo, furnished, until I sold it. The loss of my belongings wasn't hard to take in the beginning, as I learned you could live without "things" in your life. But now, I seem to feel the loss more because I identified with all my things.

How can anyone know the pain and devastation of leading a normal life and then living and feeling like a social outcast? It's certainly depressing and defeating, but I was lucky enough to have a good networking system and was able to improve by a series of moves to other areas.

I first tried a nutrition clinic in Florida, and that enabled me to get away from the isolation I had been experiencing as I avoided chemical exposures. Unfortunately, I couldn't stand the humidity in the summertime and had to leave. I ended up in the high desert of Arizona, where I improved some in the clean mountain air and was encouraged by meeting other EIs there. It was an isolated area, however, far from airports and holistic doctors, and safe, affordable housing was hard to find.

Once again through the MCS network, I learned of a clinic in Las Vegas and moved there. I found housing with an MCS couple and stayed there almost three years. But when the couple got a divorce, I lost my safe housing and could find no other place to live. I went back to Arizona, where I spent almost a year moving from one temporary place to another. Many times I slept in my car. I couldn't handle this nomadic life physically, emotionally, or financially, however, so I returned to polluted southeastern Michigan, where my aunt took me in. I've been in her not-very-safe house for eight years now.

I come from a very close-knit Italian family, but as a result of being ill and moving away, I missed out on a lot of family events that were important to me: first communions, 16th and 21st birthdays, Christmas and Easter celebrations. I have also spent my own birthday, Christmas, and Easter alone many times. I have gone through many, many dollars of my savings, and that has left me unable to afford to buy or rent a house on my own. I am at the mercy of my generous and supportive aunt.

The nice vacations, the good job, good wardrobe, and the ability to go where and when I wanted are all dim memories from the past. In fact, my past hardly seems real anymore. Some days *I* don't feel real. I feel like a stranger looking in on another stranger. The real me got lost somewhere along the way. I avoid meeting anyone from my distant past because the person they knew way back then doesn't exist anymore. That loss is almost as bad, if not worse, than the illness itself.

I resent that this illness so far seems to be untreatable and is ignored by the AMA, and I resent being unable to find proper housing because of my financial situation. It is so frustrating and a waste of a good life. At times everything seems so hopeless and futile, and then a rare good day comes along and I think perhaps there may be a way out of this nightmare after all.

I will not give up trying new things I hear about, even though the last one was a disaster for me. Perhaps something will come along in the future. In the meantime, my belief in God has strengthened as a result of the illness, and for that much at least I am grateful.

Jack

Lab Technician

In 1969, I started working in a leading research lab. All employees were given a complete physical before they began working there, and my physical indicated no problems. I was a very healthy person and had always been very active. I played basketball and baseball in high school and went hunting and fishing whenever I had a chance.

At the lab, I worked with a lot of solvents on a daily basis, particularly with trichloroethylene and freon, which were metal degreasers. I would pour 15 to 20 gallons of these solvents into open tubs or baths that I used to clean off parts for the huge accelerator. Most of the parts were made of copper or stainless steel. I would wash them with the trichloroethylene and rinse them with freon because they had to be very clean. The exhaust system above the tubs did not work properly, so I was always breathing a lot of fumes.

We were given no special clothing or respirators, although we did have plastic face shields. The gloves we were given would work for only a few days before they expanded from contact with the solvents and would tear. I later found out they were not the proper gloves for handling the solvents I was working with. And some of the screws and other parts I had to clean were so small that I couldn't handle them with gloves, so I would have to reach into the solvent tub with my bare hands.

We also used these solvents to wash out metal pipes that would be a few inches to a foot in diameter and up to 20 feet long. We would soak rags in the solvents and drag them through the pipes. When we dragged them out the end of a pipe, we would get a strong smell from the vapors.

One of our other jobs was to use aerosol sprays to coat copper parts with a clear acrylic-type finish when they came out of the furnace. We weren't given masks to wear, so of course we breathed in the mist that was floating through the air. We used to wash our hands off with ethanol all the time.

I was sent to school to learn how to repair turbo molecular pumps and cryogenic pumps and keep them running properly, so I was often asked to work on those pumps. They were installed in confined areas, and I was exposed to a lot of oil vapors in that work. The pumps generated so much heat that the section I worked in didn't have to be heated in the winter, and this heat increased the oil vapors given off by the pumps.

Another one of my jobs was to work in the furnace facility, where I would do soldering and welding. That exposed me to vapors from lead and tin. I couldn't avoid breathing the stuff. When loading the furnaces, we would use a fiber frax material to protect the parts; it was known as a "brother" to asbestos, and we could see the fibers flying through the air when we would cut it into pieces.

By 1973, about four years after I had started working at the lab, I was loaded with upper respiratory problems. I had lots of sore throats and sinus infections, and I had headaches like crazy. Even today those symptoms are still a big problem. I had never had any headaches to speak of before I went to work in the lab and hardly ever got sick; I just had an occasional cold. At any rate, when I began getting sick so often, I started seeing doctors at the lab. They treated my symptoms but never asked me anything about what materials I was working with. Even the private doctors I consulted never asked about my exposures at work. Since I was living and working in a company town, perhaps that's not surprising.

By the early 1980s, my eyes would often hurt and get very red at work; my left eye in particular would swell up. My vision would also get blurry. I would frequently become dizzy and nauseated, and sometimes I would black out for a few seconds. Around 1983, I developed terrible fatigue and insomnia. (Some nights I could only sleep for two or three hours.) I just didn't have any energy. When I told one doctor that I thought the solvents and other things I was working with might be making me sick, he said I was just imagining it.

I still didn't quite realize the extent to which the work exposures were making me sick. Then one day I was putting three gallons of freon into a tub to wash some parts, and I got a terrible headache. My eyes again became very red, and my vision started to blur. That's when I began to realize that the exposures to solvents and other chemicals in the lab were making me sick.

One day while I was working with freon, the safety committee members came through making their rounds. I asked them if I could have a proper exhaust system in my work area because the materials were making me sick. They said an adequate exhaust system would be too expensive to install at that point, but maybe there would be one at some time in the future. They told me that I must be imagining that the solvents were bothering me and said the solvents I worked with were very safe to use and would not cause any harm. I explained that I had terrible headaches, was dizzy all the time, and sometimes would even black out for an instant. When I also pointed to my red eyes and told them about my cough and sore throat, one of them said

I must just have a common cold. I replied that it was a cold that hadn't gone away since 1972.

Eventually, around 1986, I was told that I had become sensitized to the solvents I was working with and should wear a mask. In 1988 they did install an exhaust system in the section where I was working, but I was so sick by that point that nothing was going to help make the workplace tolerable for me.

By the late 1980s, I would sit up in bed almost every night at 2 or 3 A.M. and cough for fifteen minutes. I had continual drainage problems. In May 1988, my private doctor requested that I be moved to a clean working environment free of chemical exposures. When the occupational medicine department and management assigned me to another area, they sent a letter to my new department stating that I shouldn't be working with or around chemicals. But the new area I was assigned to was just as bad as the old one. My work area was in an electronics shop where people were soldering all the time, and I still worked with chemicals and with lots of diesel equipment. Finally, one day in late 1988, I was asked to work punching out holes in aluminum. I had to use a cutting fluid, and I was working in a corner with no ventilation, even though my doctor's orders had said that I was not supposed to be working around chemicals. After twenty minutes, I went to my group leader and showed him how red my eyes were getting. He said I should try using a fan to keep the vapors off, but of course that wasn't going to help much.

Finally, in December 1988, I was placed on medical leave because management was concerned I might be a hazard to myself and co-workers. I couldn't understand why they hadn't worried about that in the early 1980s when I told them I was so dizzy and had such blurry vision that I had trouble walking on the catwalks that were five feet off the floor. The letter announcing the decision said that I had a very serious medical condition, but it was treatable. The company doctor sent me to the company psychologist, who didn't ask me anything about what materials I had been working with. She just kept wanting to know if I was having family problems.

I've had a CT scan that shows I have sinus damage and a CT scan that shows very dark areas on both sides of my brain. When the technician who performed the scan looked at the film, she said it looked like I had been in a terrible car accident.

A university did a work history on me in 1989. After the work history indicating all the solvent exposure was completed, it put my mind more at ease because I at last knew what was wrong with me. Then I said, rather naively, "Now we know what's wrong with me, cure me," but they told me there was no cure. They sent me to San Francisco to consult two leading doctors in long-term exposure to solvents. Their report indicated that I was indeed suffering from solvent exposure.

While I was still working, I would sometimes lose my sense of taste and smell at work, but then it would return. Four years ago I lost my sense of taste and smell again, and it has never returned. I still suffer from a lot of joint pain in my ankles, knees, and fingers, and I continue to have respiratory problems, although they are slightly improved. Fatigue, insomnia, eye pain, and blurred vision still bother me a lot.

My health was badly damaged during the twenty years that I worked with toxic substances in that laboratory, but my former employer has done almost nothing to compensate me for the occupational illness I acquired there. I can no longer work because I am so sick, and I don't receive much money under the company pension plan because the lab forced me to retire early.

Nancy

Computer Systems Analyst

On October 3, 1998, the *Albuquerque Tribune* reported:

A life-threatening disease often forced Nancy Noren to flee her Rio Rancho home in the middle of the night to find refuge and a few hours' sleep on a remote mesa.

But it was on that mesa that the 51-year-old Rio Rancho woman may have met her death.

Authorities confirmed Friday that a partly concealed body found earlier this week west of Albuquerque was Noren, who had been reported missing since July 17.

The police eventually arrested a twenty-two-year-old man who was stopped for speeding while he was driving Nancy's car. (He had abandoned her truck in a parking lot.) Notes he had written indicated that he had been stalking Nancy on the mesa for some time. He is to be tried for her murder in February 2000.

Friends of Nancy relate how her extreme sensitivity to the pesticides and herbicides that her neighbors sprayed on their lawns affected her. One friend reported:

Nancy was totally incapacitated by pesticides and told me that it affected her whole body. The worst thing was that some of her neighbors were just not willing to talk about it at all and were quite rude about it. She wanted to ask them if they would at least warn her before applying pesticides so she could have the windows and doors closed, but they weren't even willing to do that.

When she would detect that pesticide had been sprayed, she would fly around the house closing windows as fast as possible, but it would still get in and cause her lots of pain. That's why on bad days she would have to just abandon the house and go walk on the mesa for most of the day. When the pesticide fumes were really bad, she would take her truck and camp in the camper shell on the back for a few days. Most of the time she camped out on the mesa because in a camp-

ground the charcoal lighter fluid and the cleaning products in the restrooms were a major problem for her.

Another friend wrote:

Nancy realized her home was no longer good for her to be in, but
she really couldn't conceive of moving for both financial and physical
reasons, and we all know how difficult it is for people with MCS to
find safe, affordable housing. She admitted to being overwhelmed by
the idea of having to move.

The following excerpt was taken from the testimony that Nancy gave at
the Town Hall Meeting in Santa Fe, New Mexico, in the summer of 1996.
At that event, dozens of persons with MCS testified before the Governor's
Committee on the Needs of the Handicapped. An attractive woman, Nancy
spoke with poise and self-confidence:

I used to be a normal person, healthy, athletic, and I had a real job.
I was a systems analyst. For the past 25 years of my life, I have had
symptoms basically coming from chemical sensitivity, which was the
diagnosis I got about 15 years ago. For the first five years, I was
housebound. I couldn't go any place without crumpling and not being
able to think. I couldn't remember my phone number for a couple of
years. I was just nonfunctional. I've come a long way since then by
working very hard and being lucky.

I've been involved with *Eco-Life*, which is a support group and
newsletter in Albuquerque, and through *Eco-Life* I've personally talked
to hundreds of people in the Albuquerque area who have some
chemical sensitivities and to hundreds more across the country who
want to be able to come to New Mexico because it is a cleaner area of
the United States. So there are lots and lots of us.

I now consider myself fragile—that's the word that I've finally come
up with to help me understand how to deal with the world. I can go
places, but I wear a respirator some of the time if I'm going to be some
place more than two to four hours. I can go out to eat, I can go to
movies, I can go to concerts. I can do a lot of things as long as I
never get into pesticides. As soon as I get into pesticides, I have
problems. I'm still struggling with an exposure from two years ago.
I'm incredibly pesticide sensitive.

The way I live my life with chemical sensitivity now is very different from what it's been. It keeps changing; every few years it's different. But now, every place that I want to go, I have to call, talk with the manager, and find out if and when they have used pesticides. Most places do, but sometimes a video rental store will not. Then they will get a new manager, and suddenly that person thinks that pesticides are necessary. It's very individual whether somebody is afraid of bugs. It doesn't really have to do with what kind of place it is and whether there's likely to be a bug there.

What I need, basically, is no spraying. We had a person from San Diego come and work with some of our health food stores last month, and he can get places to where they can exist with no pesticide spraying at all. I've had a very good experience using the American Disabilities Act at the University of New Mexico. They've hired an IPM coordinator, and I can be safe there.

IPM means integrated pest management, which is not what I actually want. I want no spraying, that's what my body needs. IPM is using the least toxic product; it means that you monitor, use traps, see what kind of a problem you have. And in the case of the local video store, there probably is no reason to spray. If you have a problem, you start with boric acid. But in the long run, I can't be around pesticide. I and lots of folks with chemical sensitivity have to be able to leave, so notification and posting help keep us safe. You also learn to have a little tolerance of bugs. They're part of the real environment, and if we keep spraying, they're going to win and we're going to be gone.

Janice

Executive Secretary

For seventeen years prior to the summer of 1982, I had a progressively developing career that included nearly ten years as an executive secretary and office manager, plus skill in various software programs and in creating procedure manuals for end users. I had an excellent work record with near perfect attendance and several superb letters of reference, and my social life was also very active. I was one of the leaders for the ballroom dancing movement in Ohio.

Then on June 16, 1982, I had a toxic reaction to a pesticide that was sprayed in my seventh-floor apartment to kill ants. This was the fourth application in fifteen months. Earlier applications had caused me to experience extreme fatigue and an inability to think clearly, to the point that I had to miss a day's work. This fourth spraying occurred after I had made specific requests for no further spraying, and it resulted in symptoms that were more severe and more persistent. The manager finally conceded he had sprayed in error.

About 6 P.M. on June 16, I collapsed in bed, and I awoke the following morning weak and disoriented. Shortness of breath and congestion in my lungs were major problems over the next few weeks. I lost my voice and had overwhelming fatigue. My vision was blurred; I was unable to concentrate, think clearly, or stay awake for more than a couple of hours. Prior to the pesticide application, however, my health had been ideal. My doctor's initial diagnosis was "chemical pneumonitis."

My employer placed me on full disability within the next 30 days. This full-disability status was extended to six months, and by June 1983 I was placed on permanent layoff because my symptoms did not diminish and my attempts to return to work were unsuccessful. By this time it was evident that I had respiratory damage, and I was using a portable oxygen tank to maintain mobility. The residual effects of this incident have been a total inability to maintain a job, even in a situation where an employer agreed to do anything in her power to make it possible for me to work in her office. Various attempts to work since have always resulted in another relapse.

Having had no previous problem with chemical sensitivity, I now react to car exhaust, heating fuels, printers' ink, newsprint, new carpeting, sound-proof partitions, cleaning products, floor wax, air fresheners, paints, varnishes, pesticides, odor of new clothes, detergents, fabric softeners, cigarette smoke, wood smoke, mildew, perfume, cologne, aftershave, hair spray, and almost any scented product.

Reactions are frequent and unpredictable and include extreme fatigue and weakness, periodic blurred vision, depression, loss of my voice, drowsiness, chronic indigestion, nausea, intense headaches, periodic paralysis of my right arm and leg, inability to concentrate or think clearly, shortness of breath, and a general lack of endurance. Medication, which of course is just another form of chemical exposure, relieves some of my symptoms, but often at the price of increasing others. For example, medication may control intense pain but increase weakness and fatigue.

Little has changed with my symptoms, but much has changed in my life. It is a learning process to find tolerable settings and activities. Avoiding chemicals altogether is obviously not possible, and limiting exposures is a constant challenge.

My income since 1984 has been at or below survival level. Housing has included staying with friends, house-sitting, and rented rooms. Fortunately, I recently found an efficiency apartment on the fifteenth floor of a retirement home that caters to people with various disabilities. It has been wonderful to have my own space again. I didn't think it would be possible on such limited income. The first few months were a struggle to reduce various chemical exposures in the apartment, but my efforts finally succeeded and I have a home-base "plastic bubble." I wear a face mask in the elevator and lobby and try to drive when traffic is low.

After seven years of medical tests, welfare claims, attorneys, appeals, and denials in battling with the Social Security Administration, an attorney in Cleveland, Ohio, helped me secure a small disability pension. This is a means of survival, which, when supplemented with a few hours work per week, helps me to retain a used car and a reasonable amount of mental stability. An oxygen tank helps keep me mobile, although each trip out of my room is a potential adventure. Some days I am strong enough to enjoy a little ballroom dancing with friends; most days are limited to reading, movies, jigsaw puzzles, and a small work-at-home job using a fax machine.

Life goes on. Gradually I have learned to adjust to the reality of chemical sensitivity. Most groups that I try to attend have cologne present, and my chronic cough kicks in. At least attitudes have begun to change. Ten years ago when I said I was sensitive to perfume, people thought that was really strange. Now everyone knows someone with this problem. I've given up most activities, however, since the disappointment of having to leave is worse than just staying isolated. Sometimes I think of myself as normal—until I move a bit beyond my "plastic bubble."

John

College Professor

I was healthy through four years of high school, and I was still healthy through four years of college in upstate New York. I first became sick from pesticide fogging in Gainesville, Florida, where I had moved in 1962 to attend graduate school. When I would hear the fogging truck come, I would have to leave the neighborhood and not return for 24 hours. At this time I began to have respiratory infections, inhalant allergies (to molds, pollens, danders, and other substances), and chemical sensitivities. I was totally unaware that this was the beginning of a major illness, however. I had no idea that pesticides could have such a long-term effect on health, and I had never heard of MCS. I was only thinking about finishing school and getting a job.

I was able to finish graduate school in 1968, return to New York, and work for ten years as a college English teacher. I was sick most of the time with chronic infections, however, especially in the winter when the doors and windows were shut tight and there was no fresh air to ventilate the workplace. Tobacco smoke, fragrances, carpeting, pesticides, floor waxes, and cleaning chemicals were some of the indoor pollutants I had difficulty with. In the warmer months, when I could open the windows for ventilation, I was fine.

In 1978 the college put in new carpeting in the building where I taught, painted the interior of the building with an oil-base paint, and began using a floor wax with a formaldehyde preservative in it. (This product was withdrawn from the market in 1979.) At this time 26 teachers with offices and classrooms in this building became sick, and they signed a petition protesting working conditions. This was an early case of sick building syndrome, but since no one was aware of such a possibility, response from the administration was slow.

Eventually the carpeting was removed and ventilation was improved. Most of those who had become sick were able to return to work. The chairman of my department and I, however, had become sensitized to the building and were no longer able to teach. With the degree of short-term memory loss, mental confusion, and dizziness I was experiencing, I found it almost impossible to lecture to my English classes and lead class discussions. On many occasions, I would simply have to leave the room, asking one of my better students to finish the class for the day. Finally, I was diagnosed with chemical sensitivity, immune deficiencies, and cerebral vasculitis and was given a leave of absence for disability. One doctor told me it would take me two to seven years to recover. That was twenty years ago.

Since I had always felt better outdoors in warm weather, I moved to the Southwest—Nevada for five years, New Mexico for ten years, and Arizona for the past five years. This has been beneficial in that I feel better, but I have not been able to reverse the chemical sensitivity.

The most difficult challenge of living with MCS has been finding a house I can live in without reacting to the building materials. I'm sensitive to wood, metal, and some masonry products, so there are few alternatives. I do best with adobe and stucco. I have not tried to build a nontoxic home because the price is prohibitive, and the results can be iffy. I once read a study that said the average American spends 98% of the day inside. I've reversed that proportion and spend 98% of my time outdoors, sleeping on my patio and cooking there on a hot plate. I use my house as an oversized closet, storage area, and bathroom. I've been basically living outdoors for twenty years now.

I have been told that early retirement is the American dream. Early retirement because of disability and a chronic, progressive illness is nothing but a bad dream, involving the loss of family, home, career, friends, mobility, income, and one's health—almost everything one holds precious.

Ruth

Business Executive

In the early 1940s, we lived in Odessa, Texas, where my husband got his first job as a petroleum engineer. It was wartime, and housing was scarce. We found a little rental house at the edge of town that was located several miles from the oil fields. But if the wind was right, the fumes from the gas flares that burned as surplus oil was produced wafted over our house frequently. Several times I was so sick that I would catch the train back to my tranquil East Texas, where my late physician father had many physician friends. I made the rounds of doctors. No diagnosis. I even spent several days at a clinic in Temple, Texas, but irritable colon was all they found. Back to Odessa only to get worse again. Fortunately, the oil company decided to transfer us from Odessa.

I've often wondered if there was a relationship between that early illness near the oil fields and the chronic fatigue syndrome and multiple chemical sensitivity that seemed to hit me simultaneously on April 25, 1971, when I woke up with an exhaustion like I had never known in my life. It was just overwhelming. In retrospect, I am convinced that the immediate cause of the severe illness that started that day was a tetanus booster. My mother had been highly allergic to tetanus shots and had almost died from one. Although I was 49 at this point and had had tetanus shots over the years with no adverse effects, I'm convinced that the booster shot I had in March 1971 led to the CFS and MCS that manifested themselves that day in late April by producing a disorientation that scared the daylights out of me. Then I started having recurrent three-day bouts of "flu." A month later I must have been sleeping 16 hours a day, and I also experienced extreme fatigue, brain fog, and weak vision. My joints and muscles ached so much that I had to give up driving because it was too painful for me to move my arms and legs. But never then or in the subsequent 28 years have I had headaches, thank God.

Food intolerances set in; chemical reactions set in. This was about the time fluoride was introduced into toothpaste. Wham! One trial and I was climbing the walls. There was very little literature on CFS/MCS back then, but somewhere I saw a reference suggesting negative-ion machines, probably as an antidote to "allergies." I bought a small one, but twenty minutes after starting it, I was darned near crawling up the walls.[1]

My doctor kept smirking, muttering "neurasthenia," and giving me B12 injections. He had been my doctor for years. Should have known I wasn't the hysterical type. Of course all this time, my son-in-law, a psychiatrist, convinced family and friends it was all in my head. Finally nine months after the onset of my illness, I had the strength to seek a new doctor, who hospitalized me for two or three days of tests. The results were all normal, but he did take me seriously. Unfortunately, his office was across town. I also sought the advice of a friend of the family who was an ear, nose, and throat doctor. No diagnosis, of course, but he put me on Elavil, which helped not at all.

Then I went to an allergist chum who had been president of the national association of allergists. He performed many tests. My most prominent reaction was an "allergy" to white metals. I was a widow by this point, but was still wearing my wedding and engagement rings. Shed them. But later a very strange thing happened. My dentist said I needed some teeth removed and should have a partial plate. After I had the plate put in, for three days whenever I would walk down the long hallway from my room to the rest of the house, I literally couldn't walk straight. I was almost bouncing from wall to wall in that hallway. Then through my brain fog I thought, "Ah, hah! White metal in new plate?" Yep. The dentist changed the clips on the plate to another material and no more wobbles.

My allergist had prescribed an antihistamine. Diarrhea hit. My other doctor prescribed another medication. They were definitely not compatible. The combination made me much worse. In the meanwhile, my allergist had put me on yet another medication. I was on it for years, though I'm not sure if it really helped. Then one day I had a severe panic attack while lying quietly in my own bed at home shortly after taking that pill. End of that medication.

During the early 1970s, I had to give up driving for two reasons: it was just too painful to move my arms and legs so much and brain fog caused me to overlook stop signs where I had stopped for years. I did not overlook stop lights, but I had become a danger to myself and others.

I had had to give up alcohol two months before CFS/MCS hit because hangovers were intolerable. I had been an alcoholic (beer) for years. I feel CFS/MCS saved my life. Had I continued drinking, I would have long ago died of alcoholism. So no alcohol for the past 28 years, and it's been wonderful to see the world through sober eyes again.

---

[1] Negative-ion machines give off ozone.

Very gradually my symptoms subsided until by the spring of 1974, I felt completely well, better than I had felt in years. I came to New Mexico, where my children live, to buy a condo to use as a vacation home, and I happened to slam the car door on my thumb in the hotel parking lot. I was alarmed at how my thumb looked, so I went to the emergency room. One stitch was all that was needed, and the doctor routinely asked when I had last had a tetanus booster. At the time I had not associated the last tetanus booster with the 1971 onset of CFS/MCS. Since my medical records were back in Texas, I told the New Mexico doctor that I had no idea when I had had a booster, so he gave me one. That night in the hotel room I truly thought I was dying, and I've been sick ever since. I told my interior decorator I had to go home but would return in two weeks. It took nine months for me to gather the strength to return to my new condo. I had had these plans for living in New Mexico and spending the rest of my life fishing and going to art museums and enjoying the mountains, and by golly that is not the way it turned out.

From 1974 to 1988, I could occasionally go to lunch, to a party, go shopping and to art galleries and the grocery store, although I definitely was still ill with CFS/MCS. I'm not certain what made me so much worse in 1988, but I definitely had a bad relapse. I'm now on the bed most of the day. I don't get dressed because if I had the strength to dress when I get up, I'm pretty sure I would not have the energy to undress for bed later. By 1988, I didn't even have the strength to sit at the dining room table any longer, and I still don't. I sit on the bed and eat my meals. No family dinners at all. I can't go to my children's homes. I can't have them over for dinner.

My insomnia has increased from 1988 to the present. I hate my schedule of falling asleep anywhere from 4 A.M. to 8 or 9 A.M. and sleeping until noon or 2 P.M. My muscle and joint pain and aching has definitely decreased over the years but still hits at times. Weakness, shakiness, and occasional head tremors have stayed about the same over the past 28 years, as has brain fog. I also have an odd little symptom of a leg, knee, or ankle simply giving way with no warning when I start to take a step. Every September now I get diarrhea and diverticulosis, which must be related in some way to pollens or molds that are prevalent at that time.

At one point, my gynecologist tried five new forms of estrogen on me because the one I had been taking had to be changed for some reason. The last of the five he tried was a patch. For four or five days, I suffered severe depression and suicidal thoughts. I was terrified. Had never before had anything like that. Patch stopped. Serenity restored. Doctor decided I didn't have to have estrogen.

Many foods make me sick, and chemicals definitely get me. My medic alert card is as long as a page almost of various medications that are innocuous, common medications mostly, a few antibiotics, but mostly little common things make me sick. The instructions on the card include the following: "Do not give any medication unless absolutely necessary. Sedatives act as stimulants." Of course paint and carpet bother me. I tore out all the carpet in my condo and put down tile floors. Since 1993, I can no longer tolerate wood fires or open gas flames. I dare not go to the beauty shop even if I had the energy because of the chemicals and fumes.

And the worst thing that's happened to me really is that I've been a voracious reader all my life, and I can no longer read books. I thought that books printed on acid-free paper were going to be all right because they do not cause me to experience the immediate severe stinging that other books do. But I can only read about 20 to 30 pages of even an acid-free book before I start to have a reaction, so I guess the ink is also bothering me. Recycled paper seems to be particularly bad for me. It's a very isolating feeling, frustrating, feeling a little bit out of touch, not being able to read all you want to. Thank goodness for TV and radio. I'd go crazy without that.

Editor's note: I met Ruth in New Mexico during the summer of 1998. Over the previous months, we had developed a warm phone friendship, and I was eager to see this woman with the strong Texas accent, whose vivid and feisty personality came across the phone lines. I was startled to see how tiny she was, like a small bird, with arms that looked like those of a starvation victim. Later I discovered that she weighed only 72 lbs. at that time, and that fall she dropped down to 65 lbs. for a few weeks. At the age of 77, Ruth seems to be surviving on sheer force of personality and indomitable courage, qualities she has in abundance.

Sandy

Teacher

I grew up in the Ohio River Valley near steel mills and related industries. This area was identified in the 1970s as the most polluted in the United States. My father owned and operated an 18-wheel tractor-trailer and hauled steel to manufacturing plants. Although the truck was parked well away from the house, his clothing and tools were always around. Our bedrooms were located above the garage.

Despite these early exposures, I was an extremely healthy individual. I never got the mumps, measles, chicken pox, or other illnesses, even though I was exposed to them as a child and as an adult. I never had headaches and could work for long periods without getting tired. I rarely went to any doctors, going mainly for school or employment reasons.

In 1971, I began teaching business and vocational subjects in rooms with electric typewriters, mimeograph and fluid duplicators, and an offset press. This equipment involved various chemicals and cleaning solvents. Since 1982, I have become increasingly involved with computers. In approximately 1988, I started teaching in lab situations in rooms of 25 to 30 computers. For the last five years, my assignment has been in a double-room computer lab containing up to 50 computers. The older computers have been replaced, and we now have state-of-the-art systems with Internet connections. Unfortunately, however, the new plastics, monitors, and laser printers are a source of outgassing and VOCs (volatile organic compounds). I have also measured the electromagnetic radiation with a gauss meter and know what range to stay in to avoid too much exposure.

Back in 1976 and 1977, I taught in a brand-new building in a center section with no windows. In 1982, I taught in another district, where my computer lab was located in a sublevel area in the newest section of the building. (The sidewalk was at window level.) The room was so damp that the lights would not come on until a dehumidifier had operated for a while. The computers and electronics acted up because of the moisture. In addition, our building housed handicapped students, and the sidewalk for wheelchairs was outside my windows. Buses dropping off the handicapped students idled for long periods of time (35 minutes or longer). The sloping wall on the opposite side of the sidewalk had been landscaped, and railroad ties had been added. This created a tunnel effect that channeled the exhaust fumes along the building where the air-intake vents for the computer lab were located.

After I began teaching in this sublevel lab, I started developing severe migraine headaches and sinus infections. Since I had just returned from maternity leave, I thought I must be picking up infections from the students. Other teachers who shared my room or the same short hallway experienced a variety of allergies. Frequently we had hoarseness or total loss of our voice, and many of us had sore throats, headaches, rashes, and other symptoms. When teachers were reassigned and moved to other areas, they noticed improvements in their health. For instance, one teacher who had been getting weekly allergy shots was able to discontinue them for nine months without any problems when she moved to a different area. After I was diagnosed with MCS, my allergist visited the building and interviewed the teachers and aides in the sublevel area. Five of the eight used inhalers (I did not), and most were going to allergists. One teacher was being treated by the Cleveland Clinic because her condition was so serious.

From October to December 1992, many people started complaining —five staff members, a custodian, and eighteen students. Symptoms included flushed faces, red and tearing eyes, and asthma attacks. No one in administration or maintenance took the situation seriously, however, and testing was not done until February. We were teased that we had "a computer bug." I was lectured about smells and told that not all smells are bad. The custodian was forbidden to fill out an incident report. On December 7, the levels of chemicals in the air were so high that I had a very bad reaction that lasted for weeks. This is when my MCS started.

At this time, I retained an attorney and refused to teach in that section. You can imagine the problems I've had since then. The next year I was allowed to teach in a different section of the building with the same equipment. My sensitivities slowly improved without any treatment. Then in 1994, I was moved to another school, and there my room overlooked the bus parking lot and garage. Twenty-six diesel buses warm up in the area outside my room, close to where our air intakes are located. Many bad days and reactions later I became much worse—almost a universal reactor. Perfumes and cleaning materials were really bothering me now, but so were a lot of other things. Four years previously allergy tests were negative; the same tests now showed reactions to everything. My immune system seemed out of control.

Looking back, I understand the conditions that contributed to the development of my MCS. Maybe childhood exposures to petrochemicals and pollution started the process, but I remained healthy until I spent many years in work environments with poor indoor air quality. I believe that over a period of years, low-level and intermittent exposures to molds, mildews, microorganisms, solvents, and vehicle exhaust fumes led to intolerance. When I started having headaches and sore throats, no doctor ever suggested that I might be reacting to chemicals or something in my environment. In 1986, I had my tonsils removed at the age of 36; the specialist even commented on how unusual it was for an adult with no history of problems to have this happen. The surgery did not clear up any symptoms. I still had numerous sore throats, fevers, hoarseness, and other infections. Cultures often revealed no bacterial causes.

It is amazing to me that it took me so long to notice a pattern. I often developed the severe headaches and started sinus infections when the weather was heavy, humid, and the barometer was falling. Review of my medications showed that between October and May of every year I was taking antibiotics for respiratory infections. I never made the connection, however, that something at work was making me sick.

My family is supportive. My husband has never liked fragrances or wood smoke, so he is sympathetic. Although he says he understands, he admits he doesn't fully relate to the extreme reactions I have. My son is very helpful and often warns me of unsafe conditions. He helps with cleaning projects I can't do and runs errands to help me avoid exposures. He is very careful of the friends he has over; sometimes he doesn't invite them to the house just because it is a hassle.

MCS has drastically curtailed my professional activities. I used to be very active in professional and community organizations. Recently I resigned from the last one because I am unable to attend the meetings, where so many wear perfumes. I no longer have any speaking engagements for the same reason. Two years ago I took three college workshop classes; I often had to leave early, come late, or go outside to get some fresh air. Every class was stressful because I had no control over so many factors—what cleaning products were used, who was in the room before our class, and the fragrances of others in the room.

At one time, my advice was highly regarded. I can't say that now. Many times I have felt as if I were a pawn. My teachers association used me as a bargaining tool, "You do this for us, and we'll do this for you." After months of hassling with this, I retained my own attorney and paid privately for counsel. Years later I am still fighting. Meetings with the administration were unproductive until my attorney became involved. Attempts at accommodation were slow and carried out reluctantly. Had some of the changes been implemented sooner, I'm convinced that I would not have had so many attacks and would not have gotten as bad as I did.

One positive note—because of my constant pushing, a school-wide health survey was done and areas of concern were pinpointed. A safety committee to evaluate indoor air quality was established in our last contract. Teachers in my building support me by not allowing students wearing perfume or cologne to come to my room, and the teachers themselves do not wear scented products on days they are scheduled in my lab. I did have a problem, however, after John Stossel presented the negative *20/20* program about MCS, which was unfortunately aired a second time several months later. One student who saw the show said he could wear his cologne again in my room because there was no such thing as chemical sensitivity. Of course, members of the administration also saw the show. I was in the midst of trying to get accommodations, and this television program made it that much more difficult. At least, I finally succeeded in having perfume and cologne mentioned in our student code of conduct. Deliberate use of them to make someone sick results in a ten-day suspension and possible expulsion.[1]

I am currently being treated for MCS and am noticing a slow, steady improvement. My energy level has increased, my sensitivity is no longer spreading to other substances, I do not react as quickly to things, and reactions are not as severe. I am optimistic that I will continue to improve.

My background may have increased my susceptibility to MCS, but poor environmental conditions at work played the largest part in contributing to a total overload. Each attack increased in intensity, and my sensitivity gradually spread to more and more substances. I firmly believe this condition is preventable in many cases, so let's hope that we can continue to educate others and make positive changes for the future.

---

[1] In one California school, students were deliberately pouring out whole bottles of perfume in the classroom of a chemically sensitive teacher. The administration had to install surveillance cameras to protect the teacher. A similar incident happened in a junior high school in Maine.

Jacob

College Professor

In 1989 my wife and I closed up the house we owned in Montana so we could spend the winter in a warmer climate in Arizona, where she would work as a nurse and I would teach. Unfortunately, while the climate in Arizona was indeed pleasant compared to the harsh Montana winters, we eventually learned that the apartment we had leased was being heavily pesticided without our knowledge. Our small apartment was part of a beautiful, sprawling house and grounds at the edge of a desert wilderness. The out-of-state owners, who occupied the main part of the house for a few months each year, had had the entire place routinely sprayed, indoors and out, for years. Even the lemon trees from which I was making fresh lemonade had been sprayed. But we knew nothing of this, and we never happened to be home when the three common pesticides that were routinely used on the property were being applied.

As an environmental activist and former Montana state legislator (I also taught humanities courses at the university level), I was quite well informed about chemical toxins and pesticides, or so I thought. In any case, I probably would have suspected what was making us strangely ill after we had spent some time in the Arizona house if I had only known it was being sprayed regularly.

When I arrived in Arizona at the age of 38, I was still in excellent health, as I had been all my life. The smell of strong chemicals had never bothered me, and I was using a diverse array of consumer products, just as most people do. I was able to run and hike in the desert for miles at a time. And despite my wife's bout with a fibromyalgia-like illness a few years earlier, she was healthy enough to join me on many long hikes up mountains and over the endlessly beautiful and mysterious desert. Initially, we never felt tired from these outings, but after a month or so, we both began to have sudden episodes of profound fatigue that were frequently followed by bizarre muscle spasms and mental spaciness.

I often slept on the cool floor to escape from the desert heat, and I eventually noticed that these odd symptoms would get worse whenever I did this. I also experienced a lot of nasal irritation and stuffiness that increased when I did sit-ups on the floor. At first, however, we just attributed these problems to the new, drier climate.

That first year I spent only about five months in the Arizona apartment and then returned to Montana to fulfill some teaching commitments. My symptoms at first improved when I returned to Montana, but that summer after I painted some rental properties in northern Montana, I started to get extremely sick again. I was tested for Epstein-Barr virus that summer, and the results showed I had abnormally high titers of this virus. I was diagnosed with chronic fatigue and immune deficiency syndrome, which some doctors at the time were attributing to Epstein-Barr virus. Unfortunately, this diagnosis of CFIDS disastrously led me to return to the Arizona apartment for another year.

When I returned to Arizona the second year, our landlords were building an additional room onto our apartment there. The smell of toxic glues and paints seemed overwhelming, and I began to have persistent ringing in my ears, frothy saliva, memory trouble, and badly blurred vision. But this time, because the reaction came on so quickly and my brain was so foggy, I thought I had some kind of severe flu that the chemicals were just making worse. And since my wife's symptoms didn't seem drastically worse, the flu idea seemed credible to both of us.

From what doctors later told me, there's little doubt that the ongoing pesticide exposures, especially at floor level, had sensitized me far more than my wife. The combination of these pesticide exposures and the potent neurotoxins in the remodeling products simply pushed me over the edge into full-blown chemical injury and MCS. But none of this became clear until much later, after my marriage, finances, teaching career, and health were crumbling. Eventually, I was diagnosed with "toxic encephalopathy" and immune system injury.

With a lawyer's help, I got the pesticide applicator who had treated the Arizona property to reveal that he had been spraying this property for years without placing the warning signs around the area that Arizona law requires. Amazingly, he also admitted that he had been taken to court on three previous occasions by people who claimed they had been injured by his pesticide applications. In two of those cases, he said, the manufacturer of the pesticide had agreed to settle toxic tort claims out of court, and he had been absolved of any personal responsibility. When I asked him why he continued to spray such dangerous chemicals without even warning people, he became very defensive and just hung up the phone. Unfortunately, I never pursued a legal claim against him or the pesticide manufacturer because I had no money to pay for a lawsuit.

Like thousands of other people who have been injured by poorly controlled chemicals, I developed horrendous chemical sensitivities, as well as food allergies. My symptoms included insomnia, short-term memory loss, gut troubles, ringing in my ears, and blurred vision. And before I finally assembled enough medical evidence to get on disability, I reached a state of near fury at the scores of ignorant, arrogant doctors whose increasingly corporate-influenced medical training simply finds no room for the obvious reality of chemical injury.

Being inclined toward politics and the environment, I became very focused on the subject of MCS. I talked about it too much with everyone I knew and devoured everything I could find on the subject. In the process, I got very educated about multiple chemical sensitivity and related illnesses, but getting overzealous on the topic can drive even close friends and family away pretty quickly, as most people with MCS soon learn. Some of my friends and former constituents even hinted half-jokingly that my MCS crusading might be the result of a mixed track record on my environmental bills back when I was a legislator.

The problem is, of course, that unless you have been hit with chemical sensitivity, you just can't relate to it very easily. Many people find it almost impossible to believe that a little bit of a chemical that is not hurting them could be sickening you so badly. And I'm almost embarrassed to admit now that this was exactly my reaction to the topic of MCS when I first read about it in 1985, years before I became ill. I remember thinking that these people who claimed chemicals were making them sick must be environmental fanatics. MCS seemed so implausible to me back in 1985 that I don't think I even bothered to hypothesize a possible medical explanation for it. This makes it a bit easier for me to handle people's skepticism about MCS now, but it's still very frustrating and humiliating to be disbelieved when you tell someone, for example, that the air freshener in their car is making you sick.

By 1997, after several years of carefully avoiding strong chemical fumes and possible "trigger" substances in personal care products and food, my sensitivities and symptoms had diminished almost to a pre-MCS level, or so it seemed. Strong or prolonged exposures to toluene and a few other chemicals, however, could still produce memory problems and mental confusion. At any rate, I felt well enough in 1997-98 to try going back to the Arizona apartment, after convincing my estranged wife to give things another chance. At this point her position was that since she had managed to somehow adapt to her fibromyalgia and possible MCS problems in the house, I should be able to adapt also.

I realize how foolish it may seem to some that I was even willing to attempt such a gamble. But beyond my feelings of desperation and anguish about our relationship, I was determined to resolve the constantly lingering doubts I had (despite the evidence and diagnosis) about chemicals in the Arizona apartment being the cause of my illness. I also thought the chemicals might have been dissipated by that point anyway because the pesticide applicator had agreed not to treat our apartment in the future.

Everything went very well for the first day after I arrived in Arizona after a six-year absence. My wife and I were genuinely glad to see each other, and detecting no hint of fumes or chemicals, I felt fully at ease and put the whole worry out of my head. Then about three hours after doing some sit-ups on the floor the second day, I started to get profoundly weak and sick with CFIDS-like symptoms. I fought the symptoms with everything I had and tried to hide what was happening from my wife. I told myself that I must be experiencing some kind of unconscious fear reaction, and I just tried to ride it out. But that night when I suggested removing some scented candles from one of the rooms because they might be making me "a tiny bit sick," my wife told me that the owners had just pesticided our apartment, their house, and the grounds two days before I had arrived. When I asked her why she hadn't told me this earlier, she apologized, but said she had been convinced by my reassurances that I was basically cured of MCS and she had wanted to believe it as much as I did. She also hadn't wanted to put any suggestions in my head beforehand on the chance that fear might sometimes play a role in MCS.

We both now realized, however, that fear was not playing a role in this case when my symptoms were so severe and the physical evidence was so obvious and undeniable. So despite our mutual distress, I had to leave and return to Montana rather than risk even renting a motel room or other lodging nearby that would probably also have been pesticided. I simply couldn't take yet another chance.

As I write this account in September 1999, my chemical sensitivities remain under control provided I am reasonably cautious. I'm trying to resume some limited form of teaching again in Montana, and I am struggling to build a stable life again after ten years of disabling illness and trial-and-error recovery.

Danielle

Three-Year-Old Child

When Danielle was less than a year old, we started noticing some food allergies. Nothing severe, but she would get congested and would get ear infections. These problems would disappear when we stopped giving her the offending foods, but the process started rolling faster and faster. First it was cow's milk, then soy milk, then goat's milk, rice milk, then wheat, then fruit juice, and it rolled on and on.

But we didn't really use the words "chemically sensitive" until after our neighbors did a pesticide spraying with synthetic pyrethroids for ants. Danielle woke up the next day with red welts on all her exposed skin. And then there was a neurological component, where she went into slow motion. She didn't have her balance, she wasn't speaking, and she couldn't walk straight. She would get drowsy, like the knob had been turned down to half throttle. She had just learned to speak, and after the exposure, she lost the ability to speak for a time.

Ever since that particular pesticide exposure, anytime she comes in contact with certain chemicals, especially pesticides, the same thing will happen. We'll know she has had a bad exposure when we see it come out in her skin. Once we see welts, we say, "Oh, that's what's happening." In addition, she'll get nosebleeds, real nosebleeds that go on and on, and if we have guests over, they leave the house because it's just horrifying. So we're really careful about what we expose her to. One thing we have noticed is that she'll sometimes start crying inconsolably. We won't know why exactly, though we've learned that it's usually because of a chemical exposure. While she's reacting, she can't tell us what's wrong, but she'll start wailing and wailing. Then she takes a nap and we give her a bath to wash off any chemicals that might be on her skin, and she will usually be OK and come out of it. Often when one of these reactions starts, she will just crawl in the crib by herself and go to sleep until the reaction passes.

We were in denial for quite a while about this. We thought, "All children cry, all children throw tantrums, all children have behavior difficulties." It took a long time until we observed her skin problems coupled with behavioral changes, and then we started to detect the pattern and understand that it was caused by chemical exposures.

Danielle attends nursery school in the home of a woman who is quite understanding. We have alerted her to call us if she sees Danielle wobbling when she walks, which would indicate she was reacting to something. The woman who runs the nursery school doesn't spray any pesticides now but instead uses less toxic ways of controlling insects. She has changed the soaps and laundry detergents she uses so that Danielle can attend her nursery school. We feel tremendously lucky that we found such an understanding person to care for Danielle.

We don't know what we will do when Danielle reaches the elementary school level, where we can't control what they use to mop the floors or all the spraying they do for bugs here in Arizona.

Finding a safe place for Danielle to live has been an even greater problem. We've lived in the house we're in now for about a year and a half. Before that we lived in another town, where we had a place that was located on a really nice, isolated street up in the hills, where the air was just wonderful. There were only about four or five neighbors on the whole street, but unfortunately they decided they were more interested in getting rid of the bugs than in looking after the health of our family.

It was kind of bizarre because we had invited our neighbors over when we moved into the community to explain our situation, and they had agreed to be careful with their use of pesticide. But when they thought it over, they decided they just didn't like bugs and they weren't going to put up with them. We must have gone back and forth for months discussing the issue with them. We asked them to please try to use a safer product or warn us when there was going to be pesticide spraying. Unfortunately, we found a level of hostility that was shocking. There was no dealing with those folks. The pediatrician had confirmed that the pesticides were hurting our daughter. It was breaking our hearts, it was breaking everybody's heart except for our neighbors and the pesticide companies. They increased their spraying to every two weeks. They wouldn't notify us; they repeatedly sprayed without warning us even though they could have called us to let us know. We would have clothes hanging on the line, and a truck would come and spray pesticide all over the place. All our clothes would be contaminated. We ended up throwing away mattresses that became contaminated—you can't put one of those in the washing machine. So we just had to move away in order to keep our daughter safe.

One great thing about our new community is that the company that takes care of the local park uses nontoxic weed killers so Danielle can play there.

Bob

Psychologist

When I was a child, I had headaches when my mother cleaned the floors with ammonia. Otherwise I enjoyed excellent health until 1970, when I was 36. At that time, I started experiencing symptoms of extreme fatigue and disorientation. I remember principally having to rest every minute or so while I was walking. It now appears fairly certain that the onset of this three-month ordeal was associated with the spraying of trees on the golf course where I played. The purpose was to kill the blackbirds or starlings that roosted there. We saw hundreds of dead birds, and I remember the strong chemical smell in the area. My symptoms abated that fall when I stopped playing golf. Visits to doctors proved fruitless, as they could find nothing wrong, and tests for mono proved negative.

While mono-type symptoms abated, I became subject to sinus infections, and my throat and voice became scratchy. This condition, later identified as allergic rhinitis, has never cleared up. I also began to notice that following exposure to the polluted air found in large cities, I became totally constipated and had to resort to enemas because ordinary remedies failed to relieve the condition. This pattern continues to this day.

Around 1980, I began to experience the occasional bad day, but just ascribed feelings of weakness, fatigue, dizziness, and headaches to the aging process. Retrospectively, I see that these symptoms were probably an early indication of increasing susceptibility to effects of chemical exposures.

In 1982, while I was on a five-day bicycle ride in Iowa, I had lunch next to a cornfield that I later learned had just been sprayed. Within a few minutes, my upper torso was aflame with a rash that burned horribly.

In February 1986, while living in a moldy basement apartment in Wyoming, I became very seriously ill. The doctor I consulted was mystified and settled for pneumonia as a diagnosis. My symptoms included extreme fatigue, a headache, a slight cough, some lumbar congestion, dizziness, and confusion. Antibiotics were prescribed, and I recovered very rapidly. Recovery from pneumonia is usually protracted, but within three weeks I was participating in a triathlon. Later that year, however, I became extremely sensitive to cigarette smoke, and I had to leave my job as an in-patient counselor in 1987 because of this sensitivity.

In early 1989, I again became seriously ill with acute flu-like symptoms. The diagnosis was bronchitis, but I noticed that I felt well at home but sick at work. My place of employment was a hospital where disinfectants were used extensively. After three months, I resigned due to poor health. I drove a school bus for a while but became very sick with fever, chills, vomiting, and headaches whenever I was in the bus.

As 1989 progressed, my health got worse and worse. By October, I had had a terrible headache for seven weeks, with never a letup. There was a strong metallic taste in my mouth. For several weeks my fever ranged from 103 to 105 degrees. Many other symptoms bothered me—vomiting, diarrhea, constipation, double vision and blurred vision, ringing in my ears, coughing, a foot fungus, and canker sores around my mouth. I was so weak that I could barely walk and could not negotiate steps. Crossing the street was an ordeal. I was afraid that I might die if I got much worse. I suffered muscle atrophy and weakness that persists to this day, although I am physically active again. Fortunately, I happened to contact a doctor whose wife suffers from MCS, and he diagnosed me with the condition.

By being more careful about exposures to chemicals, I recovered sufficiently to secure a position as a psychotherapist in a public health agency in Texas. The environmental medicine physician whom I saw there confirmed the diagnosis of MCS and recommended that I leave my job to find a healthier place to work. I needed the money to survive, however, so I stayed on, even though both clientele and staff smoked in all public places.

Despite the cigarette smoke at work, my condition improved to the point that I was able to attend concerts, symposia, church, and other activities. Unfortunately, in 1993 two new nurses were hired who really liked to use disinfectants, air fresheners, and potpourri. Both women were greatly concerned with ridding the air of noxious body odors and germs, and scarcely a day went by that they did not spray their offices with an aerosol disinfectant, potpourri, and air freshener. Their offices were across the hall from mine, so the fumes were constantly drifting into my office, primarily under the door.

To exacerbate matters, the mechanism for opening my window broke, and management refused to fix it because they objected to my practice of opening my window to let in a little fresh air. They insisted that the building had to be kept air-tight, by which I suppose they meant that the same contaminated air was to be recirculated over and over again. It was your classic sick building in the making.

Then one day a foul-smelling, blackish cloud poured out of the ventilation shaft. A colleague, acting on her own, had the shafts leading to her office inspected by an independent consultant, who discovered that the shafts were rife with various molds. Management, however, saw no reason to take any action in the matter except to toss a few tablets of chemicals into the vents, even though the company doing the assay recommended complete removal of the offending shafts and installation of new ones.

In 1993 there were seventeen meetings of staff with management concerning improvements in working conditions. The principal topic at every meeting was the practice of ubiquitous cigarette smoking. Fears for public health and employee health were paramount. The director, who was a chain smoker, decreed that the main floor could opt for a smoke-free ambience if it desired, but the second floor—where he had his office—would remain a smoking floor. Unfortunately, my office was located on the second floor, and it was in vain that I petitioned, begged, and implored to be relocated on the first floor. Management kept putting me off, week after week, month after month, with promises of meetings to address the problem. The meetings never took place. On one occasion an office with new paint and a newly refinished floor became vacant, and they offered it to me. I tried it out, but it was of course intolerable, and I returned to my office on the second floor.

During this period, 1993-94, my symptoms returned with a vengeance—headaches, dizziness, disorientation, fatigue, fever, indigestion, depression, sudden thoughts of suicide, and extensive liver damage. It was terrible having to go to work each day knowing how ill I would become. It took longer and longer to recover on weekends, but I did not resign until I had reached the point where I no longer could recover by the end of the weekend. All along, I hoped that management would recognize the validity of my illness and act on my behalf, but they never did.

I resigned in October 1994 and have been unemployed ever since, living first on my savings, then on the pittance I get from Social Security ($491 a month in 1994, $611 today). The Social Security Administration has treated my case—and countless others—in a cruel, capricious, and inhumane manner. I got the feeling they were hoping I would die and leave them alone. My liver is in a condition normally found only in late-stage alcoholics, those for whom death is imminent because damage is irreversible. I believe that exercise keeps me going, but I am 65 and can't keep this up forever. It bothers me greatly that our society chooses to ignore the plight of those who, for whatever reason, cannot endure the toxic conditions of contemporary post-industrial society. I feel that I have a lot to offer, but being unable to tolerate "normal" highly toxic work environments, I am relegated to the shelf.

Carole

Administrative Assistant

My doctor thinks my problems probably began about 20 years ago when I worked as an administrative assistant in a new building. The facility contained new wall-to-wall carpeting, large windows that didn't open, and a ventilation system that circulated air from various areas, including the chemistry lab, to the rest of the building. Computers were introduced at that time, and I was sent off to be trained and bring back the knowledge to the rest of my department, a state agency that, ironically enough, handled water supply and pollution control issues. I began to experience headaches, fatigue, eye strain, backaches, and muscle weakness. Office work, I concluded, was making me sick.

I decided to rent out my home and go back to college to finish my degree in something that would enable me to avoid sitting at a desk all day long because I was having some back problems. I also wanted to go into a career that would be mentally more stimulating. Then perhaps I would feel better. I headed off to Vermont with my nine-year-old daughter to help develop an experimental program for on-campus study for single parents and their children. Although I finished my studies and received my degree in psychology and theology (my passions), my health continued to give me problems. I had sought some alternative therapies to no avail and decided to take the advice of my homeopathic doctor to attend a month-long intensive healing/detox seminar in Hawaii. The sale of my home paid for this and a much-needed vehicle and gave me several months to try to recover my health.

I returned from Hawaii much healthier, but still having problems with fibroid uterine tumors. Unfortunately, shortly after my return, my rent tripled, so I accepted a job as a live-in care provider for an elderly woman and her bedridden husband, a delightful man from India with a beautiful soul. This job solved the problem of housing and work, gave me a certain amount of flexibility, and allowed me to be a stay-at-home mom for my daughter, who was now 13. I hoped I could work around my now chronic state of exhaustion.

Not long after we moved in, however, the elderly woman was diagnosed as a paranoid schizophrenic and was taken to a nursing home. Also, I could no longer avoid the fact that I needed a hysterectomy. I had the surgery, and the bedridden man with the beautiful soul was taken to a nursing home. My daughter and I stayed with a friend (thank God for friends) until I recovered from the operation and could rent an apartment. My back problems were prohibiting me from working, and my hormones were raging relentlessly in spite of medications. I applied for Social Security disability because by now my memory was giving me so much trouble I found it very difficult to work. Fortunately, it was summer, and my daughter and I moved into a healing community in the woods of Vermont, where we could spend time splashing in the brook, gardening, and eating healthy food. In spite of the poverty that was biting at our heels, we were fortunate.

We eventually landed in New Hampshire at the home of an attorney who was suffering the beginning stages of Alzheimer's disease and needed live-in help. We were just what he was looking for. It was now about two years from the time I had applied for Social Security disability, and it was finally granted, but of course, I was now working and no longer needed it. I gratefully received the lump-sum payment for past months and paid off my debt.

It was when I was living in this newly renovated country home that my chemical sensitivities were diagnosed. Less than a year after I moved in, I began to have symptoms of depression, fatigue, muscle weakness, more back pain, digestive problems, "spacy feelings," and irritability. As I have often laughingly said to my friends, I knew I was in trouble when I began to ask my Alzheimer's patients to help me remember things. And often times, remind me they would. The doctor told me to get rid of all my household cleaning products and replace them with nontoxic substitutes.

I was holding my own in terms of health until my doctor went back to Long Island. I then saw a naturopathic doctor, who recommended a good diet, good exercise, and an intensive nutritional program (which cost a fortune, but helped). He also suggested I have an affair. I followed through on the diet, exercise, and nutritional advice.

As my health deteriorated, I became convinced I was suffering from burnout. After all, it is common for caregivers to feel this way, and I was managing not only the full-time care of the lawyer, but running his huge home and taking care of my teenage daughter as well. I stayed in this position for over six years, however. My daughter graduated from high school and set off to college. Then a year later it became apparent that one person could no longer provide all the care that this man required. He left for a nursing home, and I collapsed. The struggle was over; I had fulfilled my responsibility, and I was exhausted. I rented a cabin for a couple months and did nothing but paint watercolors (a newly developed passion), eat, and rest.

To address the problem of possible burnout, I decided to do something that I was passionate about. I headed for Massachusetts to join a residential community of three women who were deeply involved in prison ministry. I soon became involved as well, going into the nearby men's prison three times a week. But the part-time job I got working in a photo shop proved to be a disaster after less than a month. I became very ill from exposure to the photographic chemicals, and while I was able to stop working by using my savings and unemployment benefits, I didn't recover for many months. It was at this time that I first noticed I had a problem with some fragrances.

The community had problems almost as soon as it started, and it disbanded within a year. I moved to another living situation with three other adults and still continued with the prison ministry, adding one morning a week at a nearby women's prison. I began to show and to sell my paintings. I worked for a while at a local farm store until I became ill from the bleach they used three rooms away in the back of the large store. Although they tried to accommodate me, it just wasn't possible, and I had to leave. Eventually I was once again offered a position working with an Alzheimer's patient. My patient was a very interesting woman who had been an art dealer. We had many lively conversations discussing art and museums.

I had addressed the issue of my problems with chemicals, and in this woman's home, it seemed, I could control all substances in my surroundings. All went well throughout the winter until the pipes burst and new wall-to-wall carpeting was installed. I haven't been the same since that fateful carpet installation. I have full-blown multiple chemical sensitivity now. I also have osteoporosis. Life as I knew it no longer exists. It is turned upside down. I can no longer go inside the prison because of fragrances and pesticides. I am too ill to work, even if there were work available that was free of all the vast number of substances that are toxic to me. My memory, concentration, and thought processes are poor. I can't be near anyone wearing any kind of fragrance on their body or wearing clothes washed in fragranced detergent or fabric softener. I am forced to live in isolation. Elimination of common household cleaners and all the products that contain fragrances has proven too difficult for some of my housemates at my rental in Massachusetts. So now I am off, like so many others with this disease, on a search for a safe place to live, watching my savings dwindle, wondering if I will be able to survive, much less accumulate any savings again.

Nevertheless, I am still very blessed. Friends of mine in New Hampshire, a family (mother, father, and teenage daughter), have invited me into their home as a safe haven. Their older daughter, who is away at college, has offered the use of her room. They have asked me to teach them about MCS and about how to live a less toxic lifestyle. They have opened not only their home, but their hearts as well. My isolation has been broken through being with people who live their lives consciously, who make their decisions based upon a calling to the highest good. Who could hope for more?

What have I learned from my experiences? I have learned about human weakness and frailty, my own most particularly. In the midst of plans and dreams being disrupted, I have learned about accepting the life you have been given and being grateful for it. The men and women I have worked with in prison have taught me to take pleasure in the smallest of things and to be straight with myself. My friends and my family have taught me about love. I have learned about forgiveness because, after all, stress makes everything worse, and living with the stress of anger and resentment can destroy one in such a delicate state of health as myself. When ill, one doesn't have a lot of options. One must pay attention, one must look for and focus upon that which is good. And there is so much good. And that is my story, so far.

Effie

Stock Clerk

I have been disabled with MCS since 1984. At least seven years before that time, I had trouble with tobacco products and smoke. I worked for a grocery chain, and when I worked tobacco and handled it through packages, my eyes would water and my throat would get sore. Smoke would make me crazy. I would fog out. I went on night-stock crew to get away from smoke half of the time. Worked with all men. Did very hard work. My life was very hard. I got colon prolapse from so much lifting. Had to have surgery.

I would go out to find my car after work and couldn't find it. It would be in the shop, and I had forgotten where it was. I would be working health and beauty aids and would get so fuzzed over in the brain I couldn't function. I usually went to breakfast right after working that department. Had 30 minutes to get across town, cook breakfast for a 90-year-old woman, eat, give her meds, and get back to work. I would find myself bearing down on the car in front of me. I was in a fog, like it wasn't real.

My heart would race and skip beats. I would hold my breath to try and regulate it. I knew something was wrong, but every doctor I went to said I was perfectly fine.

I would go to church and freeze to death. I would almost be in tears when I got out. Now I know it was reactions to perfumes and other cosmetics.

I wore four layers of clothes at work summer and winter. I thought they were trying to freeze us to death. I now know it was gas heat, pesticides, smoke, and perfumes affecting me.

People would come in the store and spray pesticides under our feet and over our head. I would run to the back room, and it would already be sprayed. Would go to the stock room and would be the same. Would finally have to go outside.

I worked every department in the store except the meat department. I'm sure I got pesticides in the produce department.

I finally got to where I could go back to church two years ago. They have made accommodations for me with special filters, and some people have quit wearing perfume. They kind of leave me alone. They sit away from me, and I feel like a lot of people dislike me.

I've had a CT scan that shows that I've had some mini-strokes. My dear husband has stayed with me through the whole thing. He is 15 years older than me and has health problems too. At least I feel like he has benefited from my diet and clean environment.

Most people think I am crazy, and anymore, I just say, "You know, I'm crazy" up front and laugh. Of course, I know I'm not crazy, but saying that breaks the ice, and they don't have to talk about me after I leave. But you know, these same people call me and want to know what to do when they have a similar problem.

Five or six years ago a very small group of MCS people got together and petitioned people to adopt a no-smoking policy in public buildings in town. We talked before a town board meeting. The officials didn't know how small our group was. We sent one man who had been a missionary to the doctors' offices and hospital. The MCS people wouldn't have had credibility. I didn't know if we had much impact, but last year a doctor's wife said to me that she thought about me every time she saw the no-smoking sign on the hospital door. Made me feel good. The manager of the store where I worked said he would call Nashville and get signs. I said, "Would you do it now and let me hear you?" He did. Progress is very difficult.

We are fighting city hall now about pesticide spraying. It's not going to be easy, but now I feel like one voice can be heard. Maybe if they think we are crazy, they won't want to fool with us. I chuckle sometimes about the whole thing. Got to find something to laugh about.

David

Psychologist

My father was an analytical chemist and an inventor. He had a lab in our house just across the hall from my bedroom, so I was exposed to a lot of chemicals as I was growing up. I remember that when I was about ten years old, I would react to these chemicals with sudden fatigue on some occasions and would have to sit down or lie down and rest. I noticed that some foods caused gastrointestinal distress and fatigue as well. In junior high and high school, I was easily distracted and was often confused about assignments. Teachers told me and my parents that I was a gifted underachiever.

Despite these developing health problems, I succeeded in getting a Ph.D. degree in psychology, and before collapsing from MCS, I was engaged in private practice and hospital work. I also devoted a significant portion of my time to evaluation and expert witness testimony in the field of forensic psychology. My wife and I had been married for 33 years and had five terrific children.

As the years went by, however, I lost ground little by little. I began to notice that I would have to suddenly sit down or lie down. The problem was diagnosed as hypoglycemia, but it didn't follow the regular pattern. Environmental exposures like cigarette smoke, car exhaust, or bug sprays would trigger episodes.

Perfumes started bothering me a lot, although I didn't find them debilitating until the late 1980s. Unfortunately, my wife loved to wear perfume and liked to have scented candles around the house. She even used one of those plug-in air fresheners that periodically spray perfume into the room. When she cleaned the house, she would put perfume on a piece of cloth inside the vacuum cleaner so that the air blowing out of the vacuum would add a fragrance to the rooms.

Of course, I found it quite overwhelming to live with this much perfume. Plagued with cluster migraines that went on for months at a time, I finally had to stop work. When I collapsed, my family said I was "going through a mid-life crisis" and "being a malingerer." They thought the condition was "all in my head" and came up with other even less complimentary or sympathetic labels. This illness places an incredible stress upon a marriage, and like the marriages of so many others with MCS, mine ended in divorce a year after I closed my practice.

I moved to a cabin we owned in the mountains an hour away to get some control over my environment. There my big adventure started. I had previously lived surrounded by people and had spent every day listening to patients and interacting with hospital staff. Instant isolation requires a very difficult adjustment, but there was no choice for me. I met a few fellow MCS people but found that the label applies to people with a wide range of symptoms. Some are simply irritated by perfume or cigarette smoke, others must live in virtual isolation. I began to learn how to take care of myself, what to avoid, and how to solve some of the problems faced by the chemically sensitive.

After the divorce was complete, my ex-wife gave me some of my old clothing, but I found that I could no longer tolerate the scented detergent and fabric softener residues on the garments. Repeated washings accomplished nothing. I had no coat I could use that first winter, which happened to be record-breaking cold. Every day I basically wore everything I owned that I could tolerate. Four shirts, two pairs of pants, lots of socks, snow boots, and a ratty old hat. Getting up in the morning was easy because I was already dressed. With temperatures frequently five or eight degrees below zero, I had to wear my clothes to bed to stay warm. Although I was using electric baseboard heaters to heat the cabin, it never got very warm, and I could not regulate my body temperature normally.

The winter of 1996-97 brought heavy wet snow that saturated the ground and filled the fields with three feet of white. I had finally gotten a settlement from the divorce and indulged in a satellite dish so I wouldn't go crazy that winter with cabin fever. That year I was alone for Thanksgiving and Christmas, as I would be for the next few years, but I still felt blessed knowing that as hard as things were, there were others who were experiencing much greater problems.

On New Year's morning that year, the wet snow became super-saturated with warm rain and slid down the mountains, damming the river, flooding the canyon, destroying homes, knocking out telephone service, and cutting off the power as the heavy poles snapped like matchsticks. Two trucks were swept into the raging waters of the river, and it would be almost a month before power was restored. The only road out was a narrow, rutted jeep trail that wound through the mountains to eventually reach the valley 65 miles away. The road was blockaded by the police, and only emergency or high-clearance vehicles with four-wheel drive were permitted to go through. The police feared that small cars like mine would get stuck and block the road completely. I had no electricity, which meant that I had no heat, no way to cook food, no water (the well pump used electricity), and no toilet. At first I thought they would get the power up in a few days. Those that had generators pumped water for neighbors to use, and they could cook and heat with their wood-burning stoves. (The smoke from a wood-burning stove was something I couldn't tolerate.)

I sat quietly in my darkened cabin waiting for some news, some hope. I began to wonder about chartering a helicopter to get out because one afternoon I walked up near the highway and saw a small plane land on the road to deliver mail. After a few days, the cabin became unbearably cold, and I was sinking into a stupor from which I realized I might not awake. I unwrapped the blankets that I was wearing, took all the money I had, my checkbook, and my two credit cards, and went up to the cafe, where I met a friend from church. As we stood talking about what could be done, two men came out of the cafe and heard us. One said that they were going into town on the back road, and he asked if they could help. When I explained my problem, they said they would try to buy a generator for me and bring it back later that day if I would give them the money and tell them where my cabin was located. Everything I had I gave to two strangers who got into a very large truck and drove out of sight in the foggy and frigid winter air.

I went home and waited. Night came and the cabin grew even colder after the sun had set. Finally at 10 o'clock I decided that they were not coming back, so I crawled into my bed, which was piled high with everything I owned that might help keep me warm, including a couple of cotton throw rugs. Shaking with cold, I tried to fall asleep. After a long time, I heard the rumble of a powerful engine, and then headlight beams flashed across the wall, sliced in narrow strips by the venetian blinds. I got up and found these two good men working by the light of their truck in the freezing cold to set up my generator and connect it to the wires I had prepared for them. Then there was the clatter of a small engine, and a light came on in the cabin. Soon I would have heat. When I tried to pay them, they refused and said they had been glad to help someone in need. I shook their hands and returned to the house.

It was clear that staying through the winter was quite dangerous, and I continued to get weaker. I tried a new medication that spring that made me much worse. By the fall of 1997, I was rapidly declining. Something had to change, so I moved to Arizona. I was planning to stay for three months, and I have been here almost two years. After the first year and further deterioration, I began to wonder if I would die here soon. At this point, a good friend of mine who had been doing some research told me she thought I probably had Lyme disease. I made an appointment with a specialist for

Lyme disease, who confirmed that diagnosis.[1] After ten months of uninterrupted treatment with combinations of heavy antibiotics, I am seeing some progress, but I am still unable to live without assistance with grocery shopping and other errands. I live in isolation still, seldom leaving the property. Life remains difficult because I have to cook outside in all weather and I am living in a small porcelain/steel camp trailer containing nothing but a bed, a bath, and a toilet. I expect that I will need to be treated for the Lyme disease for several years, but I do see little bits of progress. Most of my sensitivities have not changed much since I started the antibiotic treatment, but I now have some hope.

---

[1] Within the last couple of years, several MCS patients have told me that they have recently been diagnosed with Lyme disease. It is not clear whether Lyme disease is one route by which someone can develop chemical sensitivity or if chemical sensitivity weakens the immune system enough to make it easier to pick up infections.

Ann

Student

In my dreams I am healthy. This morning's dream was about meeting an old friend I hadn't seen since high school. The bar was hazy and loud, and our exchange fed off the chaos around us. We made our way through the packed bodies, and I introduced him to all my friends. I felt good, connected to others, and underneath it all lay the assumption that my life opportunities were like those of the healthy twenty-somethings around me. For these moments I was myself: sharp, happy, and free.

I woke, the dream faded, and the pallor of my present housebound days settled in. My bed, my room, and no schedule today, no people today. I remembered that there are lives like mine—and that mine is one of them. Now that I'm mostly housebound with severe chemical sensitivities at age 27, it still catches my breath, stuns me, and leaves me cold. I'm not that good at being sick, even after all this time. I still fight with issues of accepting it versus raging against it—especially when my health takes a serious step down like it did five months ago.

Up until then, I called my illness fibromyalgia and CFIDS (chronic fatigue immune dysfunction syndrome) because I had been diagnosed with both. Really, it just depended on which doctor I saw and whether it was the pain or the fatigue that was predominant. Whatever its name, the predisposition for it runs in my family. My grandmother has a host of physical problems resembling fibromyalgia, my mother has mild fibromyalgia and migraines, and her sister has more pronounced (but not disabling) fibromyalgia, with some chemical and food sensitivities.

I grew up relatively healthy, with some indications of the illness to come, but nothing serious. I went to college wanting to study whatever I found interesting, do well academically, travel, get out and make my life. I thought I might join the Peace Corps or Teach for America before eventually going to graduate school, but none of that would happen.

It was during my first year of college in 1991 that the problems began—frequent infections, worsening back pain from an old high school injury, concentration problems, and uncharacteristically poor grades. I didn't want to continue with school in the fall until I knew what was wrong, so I went home to work for a term. Several months later I came down with a severe case of mononucleosis and was diagnosed with fibromyalgia shortly after. My main symptom, by far, was muscle pain, mostly in my upper back, shoulders, neck, and calves. It was constant and often severe. I also had poor sleep and physical and mental fatigue.

At that time, I was living with my father and my step-mother; my room was in their basement, which had just been finished with new carpet, walls, and ceilings. I always kept my window open, even in the winter, feeling like the smell was too "strong" for me. But I never thought there might be any connection to my steadily diminishing health.

Over the next two years my pain spread and became more intense, and my fatigue became more pronounced. I finally had to quit my part-time job in a bookstore because it was making my symptoms worse. Those years were a sinking blur to me. My plans slowly slipped away, and I was very disconnected from my peers. I didn't want to face the illness and was always fighting with myself and others about just what the illness was. To lessen my isolation, I enrolled in one class at the local university, even though I had never wanted to stay in my hometown for college. Shortly thereafter, my father and my step-mother no longer let me live with them (she believed all along that it was poor character that kept me from accomplishing more).

Under my numbing disbelief was fear. I was too ill to make enough money to live on. Was my life really going this way? What happened to school and friends and internships and traveling? My world was now light-years away from all that. I tried to get a job, but no one would hire me when I revealed my physical limitations. My mother, who had little money herself, suggested that I apply for Social Security Disability Insurance/Supplemental Security Income. I protested for months because I had a hard time applying the label "disabled" to myself and couldn't believe that I wouldn't soon be better. Finally, only under duress, I started the long application process, but I eventually dropped the application when I received the second denial.

During the next five years, I worked to build a better life. The beginning was rough; I was haunted by the attitude so many people have toward fibromyalgia and CFIDS. At school I tried to hide my illness and was uncomfortable with the ways that it did show, like the way I had to constantly move and use a special table and chair to keep the muscle pain from becoming overwhelming.

Slowly facing some facts, I reapplied for Social Security, and finally, after a two-year application process, a judge did issue a "partially favorable" decision, granting me SSI (Supplemental Security Income) but not SSDI (Social Security Disability Insurance).

In the meantime, my health slowly worsened. Various injuries waxed and waned: tendonitis in both hands and arms so that for about a year I had little or no writing endurance, bursitis and tendonitis in my hips so that I needed to use a wheelchair for distances, and other less pronounced difficulties. The fatigue grew, but I made small improvements in my overall level of functioning by diligently participating in a new fibromyalgia exercise program, learning how to better manage my symptoms, and pursuing as much alternative treatment as I could afford. I had pain most of the time, but not constantly at this point. I attended forums on my illness, read, and educated myself and my doctors. I eventually made good friends who did their best to accommodate my lack of endurance and my sensitivity to their cigarette smoke. I got involved with my local independent living center to better understand the range of human experience with disability. I chased away the shame I felt about my illness, founded the first on-campus student disability organization since the early 1970s, and got to know other students with disabilities. Finally, I had reached a sort of equilibrium with the illness. I knew what had been compromised and what my limitations were, but I was also determined to make the best of what I could have. My life was starting to feel a little bit like my own. I decided to transfer to another college. I was ready, illness in tow, to move forward to new places, challenges, and a chance to build more of the life I wanted.

Then, just a week after I had sent off a transfer application, the chemical sensitivity that had been steadily increasing for a year went over the edge. A perfume exposure at the campus office where I worked a few hours each week sent me home ill. I knew it would bother me, but I thought I would be over the reaction by the end of the day. Instead I had a headache, major fatigue, increased body pain, and mental fuzziness for two days. On the third day I went back to work and had more difficulty driving because of the exhaust fumes than ever before. Then when I entered the building, the air bothered me immediately, although it never had before. My head started to ache and feel funny. When I got into the office area, someone had perfume on. I tried to wait it out and see if I could tolerate it. After about 15 minutes, however, my headache worsened until I had intense, sharp, stabbing pains that did not let up and prevented me from being able to carry on a conversation with my supervisor. I finally left. And that was it—I haven't been the same since. There were two identifiable exposures that shortly preceded my greatly increased sensitivity to chemicals. Several days earlier I had walked by a park that had just been sprayed with 2,4-D herbicide. It was also the height of the pesticide season in the farms surrounding this flat, windy city.

I'm now sensitive to many chemicals that one encounters in everyday life. My symptoms include sore throats, sinus irritation, headaches, mental fuzziness, confusion, increased body pain, severe head pain, nausea, weakness, balance problems, stinging and burning eyes, sneezing, deep coughing, nosebleeds, and heavy fatigue that lasts for days after an exposure. These reactions can arise from exposure to things like vehicle exhaust or cleaning products or other peoples' personal care products.

So began my present period of being mostly housebound. I do get out a couple of times a week, usually late at night or on weekends, when the traffic is down. I've become sensitive to the charcoal masks I use, though, so they're of limited use, and they also don't filter out enough of the chemicals in places like retail stores. I am a student living on SSI with debts and very few family resources, so I can't afford even to begin to address the host of changes I need to make to my tiny apartment in order to make it safer.

I don't yet have a physician who understands MCS. My internist was a jerk at first, smirking at my mask and sending me a scathing journal article about how MCS, fibromyalgia, CFIDS, Gulf War syndrome, and other similar conditions are psychogenic and propagated by social factors. After watching my recent decline, he is now sympathetic but has no experience with MCS.

So for now I just keep trying to get through each day. My life, though, is stopped. I'm hanging suspended above my hopes and plans that have not yet had a clear shot. I wonder if I have the resources to handle this feeling that the pieces of my life are falling away like leaves from a tree before the harshness of winter. I question how to keep being and feeling like myself when I can't give action to what I care about and want, sometimes desperately. And I miss being out with my friends—terribly. They are my context and my grounding. I do have hope for better health, but after these years of illness, I'm low on energy and more wary of looking forward to joys that may not come.

My bed. My room.
No schedule today, no people.
Is there a today?

George

Surgeon

As a child, I had many food allergies that were generally aggravated by processed foods containing preservatives. I learned rather quickly to avoid all those foods that caused me to break out in a rash in anywhere from one to three days. Although I was not necessarily rash-free from unprocessed foods, I could definitely exacerbate my symptoms significantly by eating processed foods.

In my late teens, I moved to Europe, where unprocessed foods were the norm and unsprayed vegetables and "natural" meat could be easily obtained. My allergies cleared almost completely. On occasion I could get skin allergies if I chose to expose myself to certain chemicals, but more often I would get a flare-up of conjunctivitis primarily with exposure to diesel fumes and chemicals such as toluene or other organic solvents. In my final two years of medical school, I worked in the morgue doing vascular research. There I was exposed to high levels of formaldehyde and some organic solvents. At this time, I began to have respiratory problems with asthma-like symptoms, severe skin rashes, and severe conjunctivitis. These symptoms promptly disappeared when I was away from formaldehyde or the other solvents.

I returned to the United States for five years of surgical residency followed by five years of private practice. I sometimes had skin rashes, lung problems, or conjunctivitis. In my last year of private practice, however, my symptoms became more frequent and more severe, with the conjunctivitis progressing to corneal abrasions that were totally incapacitating. I either had to take corticosteroids regularly to control my symptoms or stay away from the hospital completely. Since I believe that illness is generally not due to some intrinsic deficiency that can be rectified by taking some medication but rather is the result of environmental exposure to harmful substances or to malnutrition, I chose to leave the hospital and the medical profession completely.

Among those substances that caused almost immediate symptoms for me in the hospital setting or continue to bother me now are cleaning products containing aldehydes, nitrous oxide used for sterilization of operating room equipment, carpeting, stain-resistant fabrics, perfumes, hair sprays, toluene, and diesel fumes.

All my symptoms disappear when I am not exposed to these products, which I can best avoid by living in a late nineteenth-century lifestyle. I raise my own food for the most part. I do not have a TV, VCR, or computer, though I do have radio and electricity and use my automobile very occasionally. Hardly any supermarket products can be found in the house that I built, and it is uninsulated. I did have problems with building paper, asphalt shingles, and cement, but the symptoms were mild and easily controllable by "diluting" the exposures. I quit medicine in 1989, began building this house in 1993, and moved here in 1996. Ten years have passed since I departed from the medical profession, and in general my symptoms are much improved.

Jane

House Painter

I developed chemical sensitivities in 1986. The onset of my illness was sudden, although my occupation as a house painter may have been a contributing factor. Everything seemed to change right after I used a wood finish on my plywood floors. Shortly after that, my dog was scratching all the time, my cat was scratching all the time, and I also felt very itchy. I went to a doctor, and he suggested a lotion containing lindane, an insecticide. Possibly he thought the problem might be fleas. I got a bad reaction to the lindane, as did my dog and cat, and soon after that I started reacting to all sorts of chemicals that had never bothered me before. My animals were exposed to the same chemicals and showed many of the same symptoms, so even though the doctors at the local medical center thought my problems were psychological, I was sure our problems were physical.

My life was utterly changed. I abandoned my old polluted cabin and built a new cabin as free of chemicals as I could make it. I had to stop working and became dependent on my parents. Fortunately, they were well off. I spent most of a year lying on my bed, only stirring to shop for food, keep the wood stove going, wash dishes, and attend to survival.

Whenever I came in contact with certain substances, I developed symptoms. Most were not painful, but they were intensely uncomfortable. My eyes blurred when I held a magazine. My pelvis burned when I was in the same room as a synthetic rug. I went into a rage after I sat in chairs made of certain materials. I became dizzy and lost my balance when I wore certain shoes.

My head felt foggy, and I spent much of my time in a daze. I felt drunk, but no one seemed to notice anything wrong. I often fell asleep after I ate. I discovered that I developed symptoms when I ate a host of foods. Sugar, tea, milk, corn, mayonnaise, mustard, cheese, garlic, onions, carrots, most fruits, beef, pork, bread, spaghetti, and eggs all caused symptoms. Many symptoms were specific to a particular food. I started eating one-food meals because if I developed symptoms afterwards, it was easier to see the culprit.

I had a choice. I could migrate through states of dazedness (brain fog), lethargy, exhaustion, anxiety, and anger, with a host of bizarre symptoms or I could adopt a life of austerity. The austerity was difficult, but the payoff in feeling good was so rewarding, there was no contest. I became austere.

After three years, I learned most of the food-symptom links and chemical-symptom links. By avoiding these things, I could be pretty comfortable. The good part of this experience was an amazing feeling of empowerment. Most healthy people are afraid of Something Awful happening to their health, and they don't feel they have much control over that possibility. I used to think and talk that way myself. Now I have found that I have a huge amount of control on a daily basis over how I feel. If I had continued to go to doctors, I could have spent thousands and thousands of dollars having my eyes tested for blurring, my heart tested for irregularities, my lungs tested because of my shortness of breath, my digestive system tested for indigestion, my urinary tract tested for frequent urination, my pancreas tested for funny reactions to sugar, my ovaries tested for cessation of my periods, my skin tested for itching, and my muscles tested for soreness and weakness. Yet most of these conditions cleared within days when I found the food or substance that triggered the reaction. Those conditions that I couldn't clear by avoiding food or other substances, I could often clear with nutritional supplements. I now saw that when I wasn't having an allergic reaction, I actually felt healthier than when I had been "healthy," and I began to suspect that I was more healthy than many of the "healthy" people around me.

I also began to suspect that just as physical problems could be cleared by allergic avoidance, many psychological problems could be as well. I could trigger anger by eating corn. I could trigger paranoia by eating sugar. I could trigger anxiety by talking too long on the telephone, which had a chemical odor. I could clear all these symptoms by avoiding corn and sugar, keeping my phone conversations short, and storing the phone outside my house. I began to wonder if other people's unwarranted anger, paranoia, and anxiety, as well as hyperactivity, panic disorder, and depression were not also caused by sensitivities to foods or chemicals.

At this point I applied to a graduate school program in counseling. I figured I could spread this new idea of sensitivity to foods and chemicals as the possible cause of many psychological problems and integrate it with more traditional counseling while I learned a new way to support myself.

I did learn a great deal in graduate school in a fine program. I also discovered, however, to my anger and grief, that the mental health field is no more open to the idea of using knowledge of chemical sensitivity as a vehicle for healing than is the medical profession. Both professions are wedded to the old idea of allergies or sensitivities as symptoms of neurosis and stress.

I came to believe that I had to choose between being a traditional counselor and keeping my mouth shut about chemical sensitivity, or shouting about the power of this approach from the rooftops and becoming a pariah in the mental health profession and an object of doubt among possible clients.

Many questions remain. How can I reintegrate myself into a world that does not recognize multiple chemical sensitivity and believes that people who have it are delusional? How can I be a counselor of chemical sensitivity to people who are not interested? At this point, in a small town, I have already made sufficient noise about chemical sensitivity; I can no longer pretend to be normal. I have a reputation as a strange person. Also, after ten years of MCS, my experience has been so different from that of my peers, I sometimes have doubts about my ability to counsel "normal" people.

Over the years I have often experienced anger about how skeptics treated my MCS with contempt, doubt, and a lack of compassion. Yet behind my anger was fear and grief. Would all people treat me as that one person treated me? Would the skepticism be overcome with time and truth, or would it remain forever? The answer, ten years later, is that many people are still skeptical and righteous in their disbelief. As far as I can see, they will continue to be so for the foreseeable future. When I first got sick, I thought that my illness would be showing up all over the country and that soon doctors would accept it even if they could not cure it. Now I am resigned that without a research breakthrough, which doesn't seem likely, conventional doctors and those who trust them will continue to treat us with disbelief. I have finally accepted this state of affairs. I hate it, but I am no longer afraid of it. I have been living with it for too long.

They say you have to mourn even cigarettes when you give them up. I gave up cigarettes. I also gave up sugar, tea, bread, and coffee. I gave up mayonnaise, spaghetti, pizza, and most cheese. I lost my sense of being a normal person. I lost my ability to eat what I wanted without thinking too much about it. I lost the ability to wear normal clothes. All these losses happened because of my sensitivities to chemicals and foods. Physically, I am pretty well as long as I exercise some restraint. I am no longer as austere as I was. I can eat most anything, but not very often. I am learning to balance discipline and self-indulgence. For years I had to carry my own chair with me because so many chairs caused symptoms. Now I find that many chairs do not bother me. Yet I still need my chair for days when I am tired and don't want to take the risk and for places where I know that the chairs are made from materials that bother me.

But even though I feel well much of the time, I feel a gulf between me and the public world, which does not recognize my reality. My funny diet, funny clothes, funny house decor, funny ideas all separate me from normal people. I feel like a Vietnam veteran who has coped successfully with the hellish experience of a lifetime and upon returning home is treated with embarrassment, contempt, and indifference. How I will cope with this social challenge I do not know. I have come to give myself into the hands of God. I hope that He/She will show me the way I should go.

This year I have decided to restart the Christmas wreath business I used to have before I got sick. This time I plan to be an administrator instead of trying to do all the work myself. I can't do hard physical work anymore, but I can use my knowledge of the wreath business and the skills I learned in counseling to work with other people to coordinate their activities.

The New Age people say that when you are doing what you are meant to do, doors open for you. It feels as if wreath doors are opening for me. Everything I have done with wreaths the past month has gone like clockwork. In contrast, everything I tried to do to start up my counseling business felt difficult, frustrating, or embarrassing. So I will put my counseling business on hold and do the wreaths and see what happens. This is the next step in the healing process: returning to work is a major step in becoming reintegrated with society.

A final word. Mention of giving myself over to God will seem alien to some people. For nonbelievers, perhaps a comparable psychological state would be to Accept What Is. What will be will be. This state of mind, giving up my mission to use knowledge of chemical sensitivity to help people with mental problems and turning instead to the more mundane wreath business has been a surrender to the will of a society that doesn't want to hear about chemical sensitivity, at any rate, not from me, not now. Paradoxically, surrendering my will to the reality of social skepticism about MCS has led me to new growth and increased power. Other people working in the cause of publicizing and applying MCS theory are all in different and unique situations. I am not at all suggesting that others give up and try something else, but for me, now, that seems to be the way to go.

May all of you be healed.

Rand

Lawyer

April 1993, age 36, and suddenly I'm faced with a debilitating set of never-before-experienced symptoms, including painful stomach bloating, diarrhea, overwhelming fatigue, alcohol intolerance, strange aversions to perfume and diesel exhaust, and worst of all, a mind-stupefying confusion. Except for a broken nose, I had not seen a doctor in sixteen years. I went to one in May, I went to another in June, and then another and another and now I'm 40 years old and I've been sick for four and a half years.

I haven't worked in four and a half years. Money is a constant concern because I've exhausted my savings and live off credit cards and the generosity of my parents. I haven't socialized in four and a half years. Long-time friends have disappeared because they and potential new friends can't handle my admittedly extreme sensitivities to smells and foods and sunlight and noise and who knows what's next. I haven't played in four and a half years. I can no longer keep a cherished dog or go to a concert or joyfully join in fun and games. I haven't had sex in four and a half years—enough said. I haven't done much of anything in four and a half years except lie around a lot and try any medication, treatment, or protocol that holds promise.

I've collapsed many times in the last four and a half years. Literally collapsed, so suddenly weak and exhausted that I've fallen to the floor, unable to stand, let alone make it to the nearest sitting spot, a spot sometimes only a few feet away. And I used to be so active; I used to hike daily and regularly play basketball and I once ran a marathon. Now the mere thought of exercise exhausts me.

I was a lawyer with my own practice before I got sick. Language, working with words, whether writing or speaking, was always so easy and enjoyable for me. Now I often get lost in mid-sentence, either unable to remember what it was I was saying or unable to remember the word that describes what I do want to say. It's common that I can't remember what happened or what I was thinking ten seconds ago. Decision making is nearly impossible because I'm unable to compare options. After struggling with one option, I'll forget what I just struggled over when I move on to the next. The simplest decision—whether to go to the store or not—can reduce me to tears.

I cry a lot these days. I cry out of anger, I cry out of fear, I cry out of pure frustration, and I cry often for no apparent reason.

Being single with no dependents has been a blessing because I'm often unable to care for myself, much less anyone else. Sometimes taking a shower is the crowning achievement of my day. One doctor recommended I try a nursing home, but he didn't understand I could never tolerate the vast array of chemicals that pervade such places.

My initial problems with perfume and diesel exhaust have exploded almost exponentially. New books, new carpet, new paint, anything with phenol or formaldehyde, any pesticide or herbicide, everything from cassette tapes to car exhaust to the big bugaboo—cigarette smoke. Not just nuisance smells, these are agents that attack me with a sudden fury that can leave me totally disoriented, emotionally explosive, and utterly exhausted.

I know it makes no sense. This is a reality understood only by those who have experienced it. Those who have no such experience may or may not be sympathetic, but by and large they have their doubts. I know these doubts because when I practiced law I had a client whose case rested on whether she really suffered from chronic fatigue syndrome. I knew she was suffering, but at the same time I always felt she could pull out of it if only she would tug harder at those proverbial bootstraps.

Believe me, bootstrap yanking doesn't work. I've tugged at everything I could get my hands on, and such tugging often only makes matters worse. I've tried antibiotics, ciprofloxacin for bacteria, Flagyl for giardia, amoxicillin for intestinal fermentation syndrome. I've tried a variety of candida treatments to counter such antibiotics, a low-carbohydrate diet and caprylic acid and nystatin and Diflucan. I've tried low-dose steroids and long-term saunas and procaine injections. I've tried antidepressants—Prozac and Zoloft and Paxil and Selzone and Elavil. I've tried vitamins and herbs and a vast, expensive assortment of other supplements. I've tried fasting and fiber supplements and foul-smelling enemas. I've tried meditation and relaxation and affirmation and visualization. I've tried acceptance and rejection and most everything in between. I've tried M.D.'s, chiropractors, naturopaths, Chinese herbologists, massage therapists, psychic healers, and strict self-reliance. I've tried stuff I believe in and stuff I didn't. I've tried stuff I never dreamed I would try.

So what's wrong with me? Chronic fatigue syndrome, whatever the hell that means, has been the most frequent diagnosis. Multiple chemical sensitivity (tell me something I don't know) gets mentioned a lot, often in the same breath as CFS. Fibromyalgia describes my aching muscles and joints. Candida, systemic yeast infection, could account for all my symptoms, but so far anti-candida treatments haven't solved them. One naturopath labeled the problem mercury amalgam poisoning, but the mercury-free dentist I went to says I'm too weak to have my fillings removed. Catch-22, a so apt description of my current existence.

I would like to move to a drier climate, but I can't afford it. I'm not sure of a safe way to get there, and I don't know if I'm strong enough to handle such a move. So I live in a house wrapped in foil, foil floors, walls, and ceilings designed to keep the bad guys—chemicals, pollens, and molds— outside the living area. I don't leave the house much, only to go to the organic co-op 60 miles away, to the doctor's office 120 miles away, and to the nearby video store, where I must wear a cotton-charcoal mask. Those who know me finally realized how sick I really was when, after being blissfully without a television for twenty years, I bought one two years ago. Living in the country, I still don't have actual TV reception, but I've now watched just about every bad movie ever made. A satellite dish is tempting, but then so is suicide.

I never thought of suicide before I got sick. Now this thought is a constant companion. Sometimes suicide is a "rational" conjecture: I feel so shitty, I've felt so shitty for four and a half years, I'm not getting any better—so why not? Often, however, the thought of suicide is not only irrational but unconscious. It comes not as a result of compiling all the reasons why I should, but appears ostensibly out of the blue. However, I've learned it doesn't come out of nowhere. The suicidal feeling arises when the county has sprayed herbicide alongside the road, when I'm stuck behind the diesel exhaust of a school bus, when I have the briefest close encounter with a perfumed person. Within a half hour of these and other chemical expo- sures, it's common that I have suicidal thoughts that are not generated by any conscious thinking process.

It's scary how chemicals take over my brain. Sometimes a chemical exposure results in an absolute inability to think—I literally cannot add two plus two or spell my name. Sometimes the result is a frightening rage reaction I cannot control. Sometimes I get suicidal. Sometimes I pass out. The result is that I live in fear of people. Not just their use of pesticides or perfume, but the deodorant they wear, their hand lotion or shampoo, their clothes laundered with "fresh smelling" detergent and fabric softener, a leather jacket saturated with cigarette smoke, anything just dry cleaned or anything polyester fresh off the rack.

Due to the difficulty of avoidance, chemicals are my biggest problem, but also especially troublesome are noise, molds, and foods. Noise I counter with ear plugs and a reverse sleep schedule that has me awake at night when it's most quiet. The household foil barrier helps with molds, but moving to a drier climate is the only real solution. The foods that make me ill I've discovered through fasting followed by reintroducing a single food at a time. The list is large; it still stuns me when I dwell on it—all grains, all meats and poultry, all nuts and seeds, anything fermented, all fruits, all dairy, plus tomatoes, potatoes, carrots, garlic, and onions. I've left out some, but it's quicker to focus on what I can eat: most fish (except tuna), seaweed, olive oil, lettuce, broccoli, cauliflower, squash, and beans if cooked carefully. At least gaining weight isn't a problem.

Since I first got mysteriously sick, it's taken me a long time to figure out what makes me sick, the chemicals, the foods, etc. What I haven't been able to figure out is why I got sick. Why me? It may be genetic predisposition; alcoholism, allergies, and diabetes exist on both sides of my family. Or maybe it was my early environment; both my parents were heavy smokers, and as a child I was frequently ill. Maybe it was all the antibiotics I took for strep throats, strange flus, and possible salmonella, the diagnosis I got when I had seizures, a mini-coma, and a prolonged hospital stay as a young child. Or it could be the over-the-counter nasal spray I was addicted to in high school or the cocaine I used in my early twenties. But the fact is I never got sick from age 20 to 36. Interestingly, I lived alone this entire time. Yet even then I was extremely noise sensitive and subsequently installed formaldehyde-impregnated soundboard in the walls of my house. So maybe it was the soundboard, the particleboard shelving inside, the wafer board sheathing outside. Maybe it was the herbicides sprayed on the logging roads I hiked. Or maybe it was the trip I took to Arizona with a heavy smoker a month before I got sick. I drank many beers and ate a lot of junk food on that trip. But maybe it was my beloved dog dying a few months before I got sick or the romantic rejection from an idealized woman a month after that. Or maybe it's just a fluke. Maybe it's all in my head.

So is it predestined, genetically predetermined, purely a biochemical response to a polluted environment, physical, psychological, emotional, spiritual? All of the above? None of the above? There's no easy answer, and while I do still search for one, there simply may be no primary cause. Hence I try to focus more on what I can do to get better and what I can learn from this experience.

I've learned a helluva lot more from this than I ever did in seven years of higher education. I've been humbled, and I'm learning humility. I'm learning patience, oh so painfully I'm learning patience. I'm learning perseverance, persevering minute by minute through the hardest challenge of my life. But most of all I'm learning the lessons of suffering. I still don't know whether suffering has any existential or absolute "meaning" (the Book of Job still makes no sense to me), but the reality of suffering has most certainly hit home.

Rand's update, August 1999: I had my mercury fillings removed. No help. I moved to south central New Mexico. Big mistake. Mold has actually been a more noticeable problem here than in the Pacific Northwest. The remoteness here leaves me isolated from medical access and other accustomed amenities. I would move back to Washington if I hadn't made the big mistake of selling my house. I ignored the advice I had received during my safe housing and relocation search: Don't abandon your old residence until you're sure your new one works.

Now after six and a half years of being sick, I'm finally facing the possibility that I'm not going to get better, that my condition will only continue to deteriorate. As that possibility closes in, I remain unsure whether it is acceptable.

Nicole

Professional Ice Skater

I was born in Belgium, where I did exhibition skating when I was young. When I was 19, I heard that there were auditions in Paris for the Holiday on Ice skating company. I thought that sounded exciting, so I went to Paris to audition and was accepted. For five years I traveled all over the world with Holiday on Ice, performing in Bombay, Calcutta, Singapore, Hong Kong, Manila, and many other cities. When our plane landed in these airports, they would spray the cabin with DDT while the passengers were still inside. We had to stay in the plane and couldn't leave until they had finished spraying.

After I left Holiday on Ice, I worked for an American corporation doing soldering on an assembly line. I was breathing lead constantly, and we weren't given masks to wear. After that job, I worked in the paint and decorating store that my husband and I owned, and for three years I mixed paint.

Even though I can now see that I had some bad exposures earlier in my life, before I became sick in 1994 I had been very healthy. I used to bicycle 100 miles a week, and I would swim every day in summer and twice a week during the winter. Then in 1994 we had our whole house sprayed with pesticide by a professional exterminating company. That was a big mistake, but we didn't know any better at the time. I had found some little spots on my chest, and we thought it was possible that our dog had fleas, so we decided to have the exterminator come.

About a month after we had the house sprayed, I developed a horrible itching all over my body. It felt like ants were crawling all over me. I went to our HMO emergency room, and they gave me lindane, which is an insectide, and told me to leave it on my body for 24 hours. That was the end of me. Unfortunately, I didn't realize what was wrong with me; I didn't know about MCS yet.

From this point on, I couldn't stand my house; I would feel much worse whenever I was there. I would itch and itch and itch and just feel terrible. I had a burning sensation on the lobes of my ears and on the bottoms of my feet. I tried staying in motels and apartments, but often when I tried to stay in a motel room I would have to leave the room at 11 o'clock or midnight and go out to my car to sleep. My main problem was the itching. Unlike many people with MCS, I never had lung problems or headaches.

My husband didn't feel bad at first, but eventually he started having itching on his head and elsewhere. We finally sold the house.

By this point, I was reacting to various chemical smells in stores or other places, so I was wearing a mask to help with that problem. One day when I was wearing it another woman saw me and said, "Oh, you must have MCS." I read a book that she recommended, and I started being careful about what products I used.

I stayed up in the mountains in a place with good fresh air for two years, and I got better. Then I moved to a house where pesticide had not been used and there had been no smoking. I pulled up the carpet, put down tile, and did other things to make the house good for me. Now I feel much better. Although I still react to quite a few things, I can now go to many stores that I couldn't go to before. I can even go to restaurants that I couldn't stand to enter two years ago. I believe that if you practice avoidance and have good water, organic food, lots of rest, and fresh air, you're going to get better.

Nicole's update, September 1999: Unfortunately, I had a cancerous growth removed from under my tongue in 1998, and in May 1999 my left kidney and part of one lung were removed because of cancer.

Joy

Beautician

I worked as a beautician from 1948 until 1955, when my first child was born. Even back in those days, the chemical exposures in a beauty shop were a problem. Formaldehyde particularly bothered me. After we washed the combs and brushes in soap and water, we had to put them in a container that had formaldehyde-soaked material in it to sterilize them. I had to hold my breath when I opened the container because the formaldehyde fumes were so overpowering. We had to go through this process because we had inspectors who visited the shop once a month to see that we were following proper sterilization procedures.

We were supposed to wear rubber gloves while giving perms because the chemicals we used back then were so strong, but I didn't wear them at first because it was very hard to roll someone's hair while you were wearing rubber gloves. My hands broke out really badly after a while though and got all raw. My doctor wrapped them in gauze that I had to leave on for two weeks. So after that I had to wear gloves whenever I used the perm solutions.

One day when I was giving a perm, I got such muscle spasms in my neck that I couldn't even stand still. My husband had to take me to a doctor, and he gave me something that put me to sleep for the rest of that day and night.

Later in my life I worked as a postal employee in a post office in a large city. In those days people could still smoke at work, and the smoke would make me feel really ill.

We moved to Arizona in 1991 and found a house to rent. One day the owner notified us that someone would be coming to treat the house for termites. We didn't think anything about it at the time, but when I was looking through some of my old appointment books recently, I saw the date for the termite treatment and also noted that right after the treatment, I had to go to a clinic because I was vomiting and had diarrhea, which made me very dehydrated. They had to give me several bottles of IV solution. At the time I didn't associate that medical emergency with the termite treatment that had just been done.

In 1992 we bought a beautiful new home in the same town, and we lived there for about five years. We eventually had to sell it at a loss, however, because the copper mines in the area were illegally dumping sulfur dioxide into the community at night. There were protest meetings, but little was accomplished. The fumes came down the ravine where we lived. I could not even go outside any more, sit on the patio, or walk around the neighborhood without feeling sick, so we decided to leave.

We were in such a hurry to move so that I would feel better that we bought a home that in retrospect we know we should never have moved into. We called the builder, and he told us his homes contained very little particle-board; it was just used in the floor. The former owners swore they had never sprayed and never used scented products. They were such "nice" people that I believed them. But it turned out that I kept getting dizzy whenever I would go in the house. Thinking it was the thick new carpet making me sick, we had a subfloor put in and $9000 worth of tile. The house was only one year old, and we knew we should never buy a new home, but we were desperate.

We stayed a month in a motel trying to heat and air out the house. Finally as funds were about to give out, we moved in and found that the cabinets that looked like hardwood were hardwood only on the doors. Every drawer, closet, and cabinet was made of particleboard, even the material under the counters. The closets had been sprayed with pesticide, so I have to keep my clothes on racks in the bathroom because I cannot even open the door to the huge walk-in closet. The washer and dryer that came with the house took months to air out because the previous owners had used scented detergents and fabric softeners. I cannot open the doors to the cabinets where the detergents were kept.

One reason that we bought this house was that it is located in a beautiful spot, with state land across the road so no one can build there. There are mountains all around and beautiful sunsets. We grabbed this place in August, not thinking how it might work in cold weather. The neighbors burn wood, and the smoke comes right into our house when the wind is in our direction.

We thought that because the area was desert and no lawns, there would be no pesticide spraying. Wrong! The neighbors spray every weed that grows so it will be nice sandy landscaping. I cannot open my windows in the summer.

We get into our car and ride for hours every day because I cannot tolerate my home. They now use oxygenated gas in Arizona, so we have to head for the mountains to get clean air. As soon as we saw that I couldn't tolerate this house because of the formaldehyde gassing out of the particleboard, we immediately put it up for sale. That was a year ago, but real estate just is not moving here. After trying to find a safer place for me to live, we finally took the house off the market. We did this in fear that another place might be worse than this one. We came to this decision after looking for a safe place and finding none. At our age we are just too tired to try again. I am 70 and my husband is 73.

So each day I live in fear of my neighbors using their herbicides and pesticides, as I am bedridden for two to three weeks each time they do. Each day in this house finds me a little worse. How I wish there was safe housing somewhere for us.

I pray my new doctor in Phoenix will look at the MCS video I gave him. He laughed when I tried to explain my problem and said he had never heard that one before (MCS).

After seven years of sharing information about my MCS with my children by sending videos, books, and articles, I found out they do not believe there is any such thing. My children and their spouses are in the medical profession. My son said that he is trying to understand, but everything I have sent is anecdotal with no medical proof. My daughter believes I am being "fed" symptoms, so that I believe I am ill. She is angry because I cannot travel 3,000 miles to visit.

My husband's children are also in the medical profession and believe it is just an excuse not to visit home. They think there is no solid evidence of any such disease, especially after they watch things like the *20/20* program with John Stossel.

Their disbelief has devastated me more than the disease itself. My children led me to believe they understood, but now I know they don't. I could take most of society believing it is "all in my head," but my own children!

I thank the good Lord for a kind and understanding husband who respects me and knows how ill I get near chemicals and helps me so very much. Some day there will be an answer, I know. I only hope I live long enough to see it. In the meantime, I wish so much there was housing for those of us with MCS. My husband and I have spent so much money on this disease.

Al

Factory Machine Operator

I was a machine operator at what is called a converting house. We coated various substances like metal, plastic, or paper with other materials. For example, we produced hard disks, floppy disks, postage stamps, scotch tape, carbonless paper, and fax paper. I worked on the finishing floor, where up to six products were being produced simultaneously, creating a synergistic effect involving several VOCs. During the summer the temperature would sometimes reach 110 degrees in this former warehouse with no windows and little ventilation. When I started working for the company, however, I was in good health, as is indicated by my preemployment physical.

On May 8, 1984, paint intended for outdoor use in high traffic areas was applied indoors on the floor very close to the machine I was operating. One of the strong chemicals in the paint was toluene di-isocyanate (TDI). About two hours later, I had a rapid and pounding heartbeat and pains in my chest. I also began coughing, and my breathing was labored. Despite these symptoms, I managed to work until the end of the shift, probably about four hours. After work, I went to see my doctor, and he scheduled me for a pulmonary function test. This test resulted in a diagnosis of asthma. I didn't have any more asthma attacks in the next few months, however.

Then on January 3, 1985, they again painted the floor with the same kind of paint, and again I had an asthma attack. The doctor gave me a bronchodilator and some medication. He also told me I should find another job away from chemicals. I continued to work at the company for over a year, however, because I had a wife and five children to support. My breathing got progressively worse, though, reducing the oxygen level in my blood and causing my skin to appear ashen.

In 1986, I went to see a doctor who practiced environmental medicine, and he had a chemical assay done on my blood and my work uniforms. The tests showed high levels of toluene, xylene, dichloroethane, and methyl chloroform (TCE, or 1,1,1-trichloroethylene). My doctor told me that I had a choice: I could quit my job now and walk away or I could be carried out eventually. I chose to quit the job that I had held for sixteen years, and I filed for workers' comp for injury and disability.

In 1987 the Department of Labor awarded me benefits based on occupational asthma exacerbated by VOCs. I was also awarded a grant for vocational rehabilitation training at a local college, where I eventually received an associate's degree in computers and business.

Initially it was hard to get accommodations at the college, but then one day a professor saw me having a reaction to a chemical during an accounting class. I had gone out to the restroom, and I later learned they had just used a cleaner that contained phenol in that restroom. I was not aware of that fact at the time because some days my sense of smell just disappears. When I returned to my seat in the back of the room, I began to copy down the homework assignment from the screen at the front of the room. I started seeing double, but I wasn't having difficulty breathing, so I didn't realize at first that I had been exposed to something. When I asked the professor to focus the overhead projector, she said it was already in focus and suggested that I move to the front row. She started looking over my shoulder, and when she saw that I was putting the debit figures into the credit column, she told me to stop looking at the screen and just listen to her read off the numbers. She had been dubious about my condition until that happened, but then she believed it and helped me get accommodations at the college.

When I had an appointment with the dean of the college one day, I had to climb a staircase to get to her office. By the time I got there, I was huffing and puffing and ashen-colored. She exclaimed, "Oh, my God, is that man OK? He looks like he's having a heart attack."

I have to take an oxygen tank with me to places where I am likely to run into chemical exposures because those exposures affect my breathing. I always have the tank in the car with me because if I drive by construction sites where they are painting, or particularly if they are painting lines on the highway with paint containing isocyanates, it affects my breathing.

My sensitivity to chemicals continues to produce many symptoms, including extreme fatigue, numbness and tingling in my legs and arms, sinusitis, and rhinitis. Short-term memory loss is a problem. I often have to ask people to repeat what they have just told me because I hear their words but do not process them. After a major chemical exposure, my head sometimes jerks. Once when I was being tested for sleep apnea, they insisted that they had to use acetone to affix the probes, despite my suggestion to use tape because I was so sensitive to chemicals. During the test I had what resembled a seizure, with my whole body moving uncontrollably. My movements ripped off about half of the probes, so they switched to tape after all.

I have been fortunate compared to some people with MCS because I have succeeded in getting awards from both workers' comp and Social Security disability.

Jennifer

Student

My whole life I've been very athletic. I was a gymnast up until sixth grade, and then I was a long-distance runner for many years. I had to give that up, however, because my knees got bad. Then I became a long-distance cyclist, and I also did other sports like tennis and hockey. I was a cheerleader in high school.

When I was 18, I went off to Colorado College and had the time of my life. I absolutely loved it and had lots of friends. I was in almost every activity you could possibly do on campus, but I also studied hard because the academics are very intense at that school.

I trained to lead backpacking trips at Colorado College, and I also trained to be a rafting guide. I was the only girl among thirteen guys, so that was kind of interesting. We spent two weeks on the river, carrying very heavy equipment and portaging the rafts. In the winter I was a skier, a big-time skier. I always hung out with the guys because the girls were too slow for me. In the winter I would teach skiing, and in the summer I would lead canoe trips. I had several major injuries since I was so active, and they seemed to take a long while to heal, but I didn't stay off them. I just kept pushing through them.

In the summer after my first year of college, I was a guide up in the Boundary Waters, an area located in northern Minnesota and Canada. Kids would come up from the Twin Cities, and I would take them on canoeing trips and backpacking trips. We would canoe a lake and then portage to the next lake and eventually camp overnight. Trips would last about two weeks. Many of the kids had no experience at all, so I would end up doing a lot of the work. We would have to carry canoes weighing up to 75-80 pounds and backpacks weighing up to 90 pounds, so it was hard and demanding work. We were pretty tired by the end of a day.

That summer I got a groin injury while I was biking in Minneapolis. When I went off to the Boundary Waters, I tried to ignore the injury at first, but I ended up on crutches. Initially, I just stayed at the base camp, cooking meals and making up backpacks, but the injury started to get worse and I was in pain all the time. As a result I had to go home to Minneapolis and was unable to return to college that fall.

Before long the injury was so painful that I couldn't sit or walk on it. I was bedridden most of the day. We didn't know why it was taking so long for this groin injury to heal. The doctors put me through a lot of physical rehabilitation and sent me to pain clinics to try to help control the pain. By the fall, I was in such bad shape that I was in a wheelchair and had to be wheeled into a handicapped pool to do range-of-motion exercises in the water. By the spring, when I was in bed most of the time, I would sit up for meals and sometimes I would try to sit up and draw for a half hour. In the morning I would get up and walk from my bed down to the couch in the living room. My college friends, who all knew how active I had been before, found it hard to believe I was reduced to lying on the couch all day.

That spring, when it was still pretty cold in Minnesota, my family painted the whole inside of our house with oil-based paints. All the windows were closed because it was still cold, so I was just lying in the fumes all day long. The fumes were pretty strong, and my parents and brother were getting headaches, but we didn't think anything about it. Then about a month after we painted the inside of the house, all the tendons in my body became really tight and inflamed and they curled up.

It was particularly devastating to have the tendon problem develop because every month I had been hoping to get back to Colorado College and all my friends. I was scheduled to go back that summer on a limited basis, but then I got a neck strain, and I was in so much pain that I had to wear a neck brace all the time and I could hardly talk. I couldn't even put my head up off the couch. I also started having a severe problem with my Achilles tendons; it always seemed to be connective tissue that gave me problems. As time went by, my tendons seemed to get even worse.

My mother had to take care of me constantly. She had to cook meals for me, and at times when my wrists were very painful, she had to feed me. I spent my days reading and watching TV, but then my eyes got so bad that I couldn't even read and I had to listen to books on tape.

Fortunately, a family friend named Marilyn realized that my mother needed a vacation for a while and offered to let me stay at their house for a couple of weeks. I only went to get off the same couch I had been on for a year, but in two weeks I started to feel better. I was able to walk. I had to walk slowly, but I could walk. I could actually go up and down stairs a little, although I had to use a cane for the stairs.

Marilyn happened to have chemical sensitivities, so she used natural cleaning and personal care products, and she had air filters in her house. When Marilyn first told me a little about her chemical sensitivities, I didn't believe that anything like that could exist. I didn't think that could be what was causing my health to deteriorate.

As it turned out, I ended up staying with Marilyn and her husband for eight months because I was gradually getting better there. I could walk better, and I was doing better in physical rehab. Whenever I would try to go back home, however, I would get worse immediately. Not only did my house still have lingering low levels of paint fumes, it was also an older home with a moldy basement and a musty smell throughout the house. I later learned that almost everyone with MCS reacts to mycotoxins, which are chemicals given off by mold. At any rate, by this point it was clear to me and my family that environment was having a big effect on my health, so I began doing some research into multiple chemical sensitivity and started avoiding products that contained chemicals that could bother me.

Then after improving for eight months while I lived with Marilyn, I got worse because of two large exposures there. A neighbor treated his lawn with pesticide, and Marilyn had a houseguest come to visit who wore heavy cologne. I suddenly developed severe fatigue all the time, and I was diagnosed with mononucleosis. My muscles had been getting better, but now I was back in bed. I ended up completely bedridden for about four months. I couldn't read or watch TV, and I couldn't concentrate enough to put sentences together.

By July, my doctor suggested that I should get away from the pollens and mold of Minnesota and go to the Southwest, where he thought I would feel better. Following his advice, my mother and I went out to the mountains of Colorado to see if we could find a place for me to live there where I could regain my health.

When we first arrived in Colorado, we stayed for a couple of weeks with an MCS friend named Jane and then camped out for a while. I feel great when I live outside even though I can't do that for long, and camping only works well if you're not breathing smoke from someone's campfire. So I began the inevitable house hunt that most people with MCS go through as they search for housing that they can tolerate. It's very hard to find a place to live when you need to avoid relatively new carpet that's offgassing toxic chemicals, old carpet that's musty, gas stoves, hot air heat, particleboard building materials, and recently painted places for a starter.

I finally settled for the winter in a cabin that was located very near Rocky Mountain National Park and was fairly well separated from neighbors with smoke from fireplaces or wood stoves. It turned out to be a real hassle when winter snowstorms arrived though. I often couldn't get my car out of the driveway. And I was tired of having to shower any place but my own house because the well water was brown and smelly.

Excerpts from a letter written by Jennifer's mother

Coming home was really a strange experience after five months away. Jen seems to be managing OK on her own, although she seems to have weekly disasters occurring all the time. I usually pick up the phone to call her with great trepidation![1] When her staph was so bad, I was sure I was going to get that emergency call to come back, so I rushed around here tying up loose ends. Andy, her boyfriend, must be a saint to do so much to help her. Without him, I'm sure I would have to go back. Saturday he went up and did all her laundry that had been accumulating for weeks. . . .

I admit to being somewhat at loose ends lately. I have been so used to constantly putting out someone else's fire. All but one of my friends are out working, so I have no one to hang around with anymore when I want a break from housework or paperwork. It's amazing how everyone else's life went on where mine just stopped for three years. I am not sure how I can resume my computer classes without starting over, as each software program has had at least three upgrades since I began the program. Lee is hinting about my joining the workforce as our savings are about depleted from Jen's medical expenses. I did send for an application for a job reading student essays for three months. I can't get tied down to a permanent job when I am sure Jen will need help in May getting relocated again. So I am unsure just what I will do until then.

Jennifer's letter to friends

Let's see . . . it all began last Memorial weekend [1995] when I had to move out of my cabin. Remember that idyllic little cabin nestled on all those acres of forest that I was renting last year? Well, I was definitely ready to leave that house! Sure it was gorgeous, and I loved all the animals around, but my health steadily declined over that year. So I opted to not sign the lease for the next fall.

My parents came out for the long weekend to help move me out again. It was quite a chore ripping all that foil down, hauling all my stuff to

---

[1] One such disaster occurred when the sewer pipes in the mountain cabin Jennifer was renting froze, causing sewage to come up through all the drains in the cabin. Jennifer resorted to cleaning up the mess with chlorine bleach since she had to use some sort of a disinfectant and apparently didn't want to buy the more expensive hydrogen peroxide, which is much safer for the chemically sensitive.

storage, and putting all the owners' furniture back.[2] My parents were sure glad to go back to Minnesota after that, and my dad vowed never to move me again! This left me without a roof over my head once again. Little did I realize it would be five months until I had a place of my own!

While moving out of my old house, I had severely strained my groin again and could barely walk. I stayed with an MCS friend, Laura, for a few weeks because my groin was so painful that I could not drive to get to the house in the mountains near Denver where Jane had offered to rent me a room. But I finally made it to Jane's. I was excited to finally be able to relax and literally "put my feet up." Since it was around the last week of June, the rain finally cleared and the weather began to warm up. Yeah! Just in time for me to camp out! You see, Jane's house is so moldy I can only be in it for short periods of time. When I stayed there the first summer I was in Colorado, my muscles got really bad. Alas, I was back again, but only with the intention of staying a short while. I set up my cot and sleeping bag (little did I know I would be sleeping on this uncomfortable contraption until winter) on the outside deck under the eave of Jane's house in hopes that the rain would come from the other direction. I still do not have a tent I tolerate. I used a corner of Jane's kitchen to place my belongings, and began to unwind emotionally in the peaceful and loving atmosphere of Jane's house.

Unfortunately, the pesticide season began to roll around. In Colorado (and most states, unbeknownst to the citizens), the state is required to spray all the roadsides for "noxious" weeds. Anyway, one day when my leg was extremely sore, we realized they had just sprayed a nearby road. If the wind is blowing from the direction of the treated road, I will feel ill even if it is three miles away. We sealed up the house, I moved in all my belongings, and we prayed that would be enough. But the spray came into the house through the cracks, so we had to "evacuate." Jane left for the day, and I attempted to stay since I was too sore to drive. By the next day, however, I had to leave, so I drove all around trying to find some place I could set up my cot and read for the day. With no luck after several hours of driving, I ended up at the house of another MCS friend and sat in her driveway for the afternoon. At this point I was beginning to feel rather unsettled and sick of all this moving around. Little did I realize this part was a walk in the park! After several days of doing this, Jane and I decided to haul our bedding to the house of another friend, who lived in a nearby valley, and we stayed there for a few days. I set up my beloved cot in her backyard and attempted

---

[2] People with MCS often cover walls or floors with aluminum foil, which acts as a barrier to the fumes emanating from plywood or particleboard.

to avoid the rain under the eave, much to no avail. Despite feeling utterly homeless, cold, and wet, I still remember looking up at the stars at night, gaining the strength and will to keep plugging along.

Jane and I finally got to return to her house, but I had to stay inside the house for the next few weeks. This is when my muscles began to be seriously affected. I began to pull my muscles for no reason with the slightest of movements. I became increasingly immobile. Then one day I severely strained my neck and esophagus. It was so painful that I could not hold up my head or even speak. I had to remain on my cot totally immobile twenty-four hours a day. I had since moved my cot outside on the deck again, so I spent the days reading (literally from 8 A.M. until bedtime) and meditating. In the two weeks that I could not move, all I had was myself and the trees around me. Plenty of time to contemplate and just be.

It was about this time that Andy came and rescued me. We took off to go camping up near Breckenridge. He would climb a few fourteeners while I waited at the trailhead. The first night we got a late start, and we didn't get to the campsite until 10 P.M. Actually, it wasn't really a campsite. We had kept driving until we could go no further, which placed us way above tree-line in the snow and a boulder field. It was sleeting and pitch dark as we tried to set up the tent. Since I do not tolerate tents, I slept with my legs in the tent and my upper body outside in the snow and sleet with my back on all the rocks. Boy, was I excited! This was the first time since college that I had been camping at that high of an altitude. What fresh air there was!

The next four days were a wonderful respite for the two of us. In the mornings, Andy would climb a fourteener or two, and then we would spend the days riding around Breckenridge. We always managed to find the most idyllic spot to camp for the night way off the beaten track. It sure was hard, though, to have to wait at the bottom while Andy made the trek up. I cannot wait till I can hike again.

After that great vacation, I was recharged, which was good since that is when it all hit! Just as I was getting ready to leave Jane's house to go back up to the mountains to look for a place of my own, her neighbor mosquito-fogged his yard for a party. Since I had no closet in which to put my belongings, they were all out on her porch. This meant that all my belongings now had pesticide on them. I was furious! I quickly began to try to salvage what I could and wash a few necessities (like my sleeping bag) and madly pack my car. In my haste as I was furiously scrubbing my Thermarest in the tub, I slipped and did the splits. I strained both of my groins this time. I screamed bloody murder, and Jane came running to help me to the couch. She stayed nearby to get anything I needed as I lay there crying for hours. The floodgates opened, and I cried for all the difficulties I had been through. Why couldn't God give me a break? I couldn't understand why life had to be so hard.

After several days of rest, I still could not walk or drive. I knew I had to get out of Jane's house if I wanted my muscles to feel better, so I swallowed my pride and called home for help. That was extremely difficult for me to do since my folks are always trying to get me to come home. Well, boy, did I get it! My family was furious! I had no choice but to come home, they told me. My mother was sick and tired of having to rescue me clear across the country!

Well, my mother did in fact fly out later that week to help me. The next morning my mother and I drove up to Estes Park, arriving quite late. We set up camp in a campground in Rocky Mountain National Park. I was excited to be back camping in the mountains, even if it was car camping. You have to understand just how much my mother hates camping, so it was definitely a sacrifice on her part.

Time for bed at last, but alas, there would be no rest for us yet. The campfires were beginning to be lit. Now I can handle a few fires if I am downwind from them, but this campground was in a bowl and it was filling up with smoke. We packed up my cot and bedding and thought we would just go lie on the other side of the grounds for a while. But that also soon filled up with smoke. My mom had to pack up our stuff again (I couldn't do much because of my muscle problems), and we drove away from the campground a little ways to wait until the campfires might die down. That never happened, however, so at 11 P.M., we drove back into town and set up my cot in Laura's front yard.[3] Then my mom had to drive back to the campground to sleep in the tent we had set up earlier. Despite all the remarks she was making about everything she was having to do for me, I still knew she would do anything to get me well and that she loved me dearly.

The next day I was elated to finally be able to lie on my cot all day for the first time since I did the splits and injured my groin. Boy, did I need that, and so did my mother. But around noon as I was blissfully reading a book in the yard, Laura came running out to tell me the next-door neighbors were staining their house. Ughhh! We packed up my stuff and decided to drive to Meeker Park (a town about 20 minutes away higher up in the mountains) to check out the campgrounds up there. And guess what? We found a beautiful site at the base of Meeker Mountain in the little rinky-dink national forest campground. I was so excited I could hardly contain myself.

---

[3] Laura was herself very sensitive to chemicals and reacted to some of Jennifer's things, so she would feel sick if Jennifer spent too much time in her house.

It was near the back of the grounds on a hill with the wind behind us so the other fires would not be too much of a problem.

Once we were all settled in, I proceeded to spend the next few days relaxing. What sheer bliss! The sun was shining, the weather was dry, and I was in the most gorgeous area I could have asked for. Yes, I knew God was watching out for me. I was in heaven with the squirrels running everywhere around us (and stealing our food when given the chance), but my mother saw it quite differently. She absolutely hated it.

My mother was using the tent, and I was sleeping on my cot out in the open. As the wonderful weather continued, I got cocky and never set up the tarp because it hindered the beautiful view of the stars. But many nights it would suddenly pour rain with torrential winds blowing everything over. Mom would come racing out of the tent, and we would proceed to scramble to stake down the tarp above my cot. The tarp of course would always blow over anyway, and I would remain soaking wet, huddled under my sleeping bag in fear of blowing away.

After a couple of weeks of camping, my mother could no longer stand it. She wanted a shower, electricity, a stove, and a phone. I must admit it was difficult to call around for housing when the realtors could not call us back. She had had enough, so she began to hunt for a temporary place for her to stay. My mother did find a few condos to rent for the month at $1000, but when she came up to Meeker Park to tell me about them so we could decide, she promptly lost both. She was devastated and at her cracking point. No more! She was going home—this became her favorite statement throughout the summer. Fortunately, after a few extremely tense and depressing days, my mom found a cabin down the road to rent for $50 a night. It was the cutest thing! Set back on the end of a dirt road, it was an old log fishing cabin with the bark still on the logs. It wasn't winterized, however, and would be closed come Labor Day. But there was plenty of time to find a house for me, or so we thought.

While my mom stayed in her little cottage, I continued to stay at the same campsite. The nice elderly couple who were in charge bent the rules and let me stay there as long as I needed. What a blessing! Let me tell you, I thanked God for each and every blessing I got. It made the tough times seem not so dismal. I would have stayed in the cabin with my mom, but unfortunately they put moth balls in the place during the winter, so that made it impossible for me to stay there. But after my mom spent about a week of traveling back and forth from her cabin to my campsite to bring me food and other supplies, it got too hard on her. We called it "meals on wheels" since we could not leave the cooler sitting in the campsite because of the bears. Plus, all my clothes and belongings were in the trunk of my car. So I decided to try setting up my stuff in the yard behind the cabin.

This worked well for a few days, but then one day after house hunting all day, we arrived home late at night, and I decided to take a shower. I noticed a strong chlorine smell from the water, but I blew it off, thinking that I was just overly sensitive and rundown after months of stress. Finally, I asked my mother to come and smell the water, and she was aghast at how strong the odor was. I immediately got out of the shower, but I already smelled as if I had dumped a bottle of chlorine bleach all over myself. My mother drove down to the front lodge to see what had happened, and they told her they had just cleaned the pipes that afternoon. Of course, they saw nothing wrong with the amount of chlorine they put in, even though others had also complained. Completely exhausted, all I could think about was how wonderful my cot would feel, but now I had to figure out how to get the chlorine off of me. I couldn't expose Laura to the chlorine odor because she was so chemically sensitive, and all the public showers were closed at 10 P.M. in Estes. Then I remembered a massage therapist I had seen once in a while in Estes who had said I could shower at her place anytime I needed. I hoped that meant late at night on a Friday evening. So we showed up on her front porch and sheepishly asked if I could shower. She graciously opened her house to us, and while I took a hot bath, my mom sat and watched a Disney flick and ate cookies with her kids. How kind! I even got to see the end of the film while I waited for my hair to dry. Since by this time in August it had gotten fairly chilly out at night in the mountains, it was such a luxury to be inside a warm house watching TV. Anyway, it was a lovely evening. I have met more kind people than I can count. The world really hasn't gone all bad.

Only a few days later we faced yet another problem. A neighbor who had a house next to our little cabin decided to stain the outside of his house. Lucky for me, Andy was coming up that afternoon, so we could escape and camp somewhere else, which is what we had planned anyway. The previous weekend Andy had found a campsite ten miles down a forest access road, so that is where we headed. We set up camp and spent a wonderful evening lying on the ground looking up at the stars. He, of course, had to retire to his truck once the mosquitoes began to eat us alive! (I could not tolerate insect repellant on him or me.) The next day he had to leave to go back to Fort Collins. I decided to stay put since I was sick of moving every day. I felt safe in this campground because there were four other occupied sites nearby. I had fun reading, meditating, and reading some more despite the constant rain. But by evening, the other people began to leave, and once nightfall came, I was left to myself. Uneasiness set in when random cars of noisy men would drive by and shine their lights into my tarp. I felt like a sitting duck, but I made it through a restless night and thought I was home free. It was 7 A.M., and my mom would be there later in the morning to pick me up.

As I began to break camp, however, a campground worker pulled up in his truck. Immediately, an alarm went off in my mind, warning me, although I am normally naive to that kind of situation. I usually innocently believe that strange men would never harm me, but this time it was different. This man was looking around the nearby, deserted campsites to see if anyone was there. Then he ambled on over to me and asked me if I had seen the owners of the empty tent next door. I naively said no and answered his questions about whether I was all alone. He was a strange character. His demeanor resembled that of Lennie in Hemingway's "Of Mice and Men." He slurred his speech, and I could tell there was very little upstairs. Finally, after haggling me with tons of questions, he got into his car and drove down the road. By then I was physically so scared that I was about to throw up, and I was shaking uncontrollably. It was not over yet though because he merely decided to go by the other sites to see if anyone was around. When satisfied we were totally alone, he immediately drove back. He sauntered over to "see if I needed anything" and "to chat some more." I really thought that was it. He knew I was alone, had no car, and was injured, so I could not run away. I managed to stay calm while he looked me up and down with a strange look in his eyes. I could tell he was debating if he wanted to rape me or not. After a short while, I guess he decided against it. I thanked God as he began to finally drive away. He parked down the hill, however, and just waited for a long time for no reason, apparently still trying to decide what to do. By this time, it was late morning, and I prayed for my mother to come soon. I tried to hike up a steep hill to hide in a deserted tent, but could not make it. Instead, I waited on the road. When my mother came, I told her nothing so she wouldn't worry. I was afraid she would not let me camp alone anymore, but I knew she was suspicious since I was still shaking and I was in the worst mood. I tried to go on with my day as usual, but I would jump at the sight of any red pick-up or a scruffy-looking man.

Later that day after looking at several more rentals, I was back to the homeless situation. I tried to stay at my mom's cabin again, but still could not be there because of the stain being applied next door. I then tried to set up camp at the old "homestead" where I had camped before, but the place was completely deserted. Not a soul was to be seen. I did not feel safe anymore camping on my own with no one around, so I decided to set up my cot in the campsite right near the road. My mom began to get very suspicious at this point, but said nothing. She came back around dinner-time, and I decided to move to a different campground down the road where there was a resident assistant in charge. We drove around endlessly trying to find the right spot—far away from other campfires, yet close enough so I felt safe. I finally got situated and just wanted to crash. By this time my mother knew something had happened the night before, so I filled her in. To make light of the situation, she said it was a good experience for me since now I might use some caution and common sense. She was right in that aspect.

So I settled in, ready for a good night's sleep. Wrong! At around midnight, I was abruptly awakened by a bear throwing my cooler around just ten feet away. I was scared shitless! The bear was having quite a difficult time trying to get the cooler open and made an awful racket. It went on for about an hour, while I continuously chanted every prayer I knew, no matter what belief or situation they pertained to. I said Hail Mary's, pieces of Jewish prayers I knew, the Lord's Prayer, New Age chants; I even tried talking to the bear's spirit! One of these has got to work, I thought. I huddled under my sleeping bag (remember, I couldn't use a tent), sweating profusely from the stifling air and anxiety. I thought I would suffocate. At one point, I could have sworn the bear was at the head of my cot from the sounds of his breathing and his footsteps. At last the bear decided it had had enough and left me in peace.

Early the next morning, I awoke to investigate the scene of the crime. My cooler remained intact, giving me the idea of writing to Coleman to tell them what a great commercial it would be to have a bear throwing their product around to prove how durable it is! Everything else was a complete mess. My vitamins had huge teeth marks gorged into the sides of the plastic bottles, and my glass jars of food had greasy saliva all over them. I laughed at the thought of how frustrated the bear must have been trying to get at the food he could see and smell through the glass but not get to. I know what you are all thinking. Why on earth did I leave a cooler out on the table for the picking? Believe me, I was berating myself all night for being so neglectful. I of all people knew better than to do that. I used to preach to groups on the art of hanging your food from a tree limb all the time when I was leading trips in the Boundary Waters.

By late morning, my mother had not shown up yet. I was famished, so I sifted through the mess and pulled out a jar with some very warm vegetables in it. I poured some olive oil over them to cover up the fermenting taste, squeezed my hand into the jar, and came up with a fistful of food. Oil and mashed veggies were dripping down my arm, and my face was smeared with them as well. But I didn't care—I was starving. By the time my mom showed up at 2 P.M. and got me some better food, it was time again to find another camping site. By this time I was tired and frustrated at not having a safe place to sleep each night. "Why, God? I am not asking for much. Only a place to put my cot each night. I'm not even asking for a house anymore." I usually like the transient life of constantly being in different areas. I get bored with the same old scenery. But all I wanted right then was consistency. I felt so ungrounded and scattered. I was sick of camping, sick of living out of my car, and sick of eating cold food out of jars. So we half-heartedly drove around the campground looking for another place to set up camp. I did not want to stay in the same spot in case the bear came back for more that night. I wanted and needed some rest. We finally picked a nice, remote spot up on a hill. My mom hauled all my stuff up, which took several trips, and I sank down onto my sleeping bag for some R&R. But just then a whole truckload of men pulled up, beer kegs and all, and set up camp right next to me. Just my luck! A huge empty campground, and they have to pick the campsite right on top of mine. After several ogles and jeers, I decided I did not feel safe camping alone next to a bunch of partying men.

We packed up my belongings once again—no easy feat with a small car—and searched for another place for me to camp. We were both frustrated, weary, and ready for that luxury suite with room service and a jacuzzi. I did get situated in another site and vowed not to move for anything. My mom left, and I had a nice, uneventful evening.

We spent the next day going to appointments and realtors in Boulder. I got back late at night and just wanted to crawl in bed. Wrong again! Since it was Labor Day weekend, the campground was totally filled and I happened to be surrounded by blazing fires and kerosene lanterns. I said, "Fuck it, I refuse to move!" But after a while it was obvious I was going to have to. But where could I go? I had exhausted all possibilities but Laura's place—yes, I would camp in her yard until the busy weekend was over. Once again we packed up my stuff in the dark and headed for Estes Park. I got situated in her front yard and fell fast asleep.

Our next immediate dilemma was where my mom would now stay because her little cabin would be closed in two days. The day before, a realtor had told my mother she could rent a condo they had for the month, but when we went there on Monday morning to sign the contract, she had given it to someone else. My mom burst into tears. We then headed on to an appointment we had scheduled with another realtor to look at a condo I was sure I wanted to sign a lease on. Andy and I had gone to look at it that weekend, and despite the hundreds of old beer bottles, moldy pizza boxes, dried-up food on the stove, and stale marijuana permeating the air—it was the epitome of a rundown college house—I thought I could fix it up. After I thoroughly looked over the place again, I excitedly ran out to say I would take it, but my mom looked at me with tears in her eyes again. While I had been inside, the owners had come by to say that in two weeks they would be spraying fire-retardant on the outside of all the units and spraying the trees for mistletoe. I was crushed. I had finally found a place I thought I could tolerate, and it was taken out from under me.

Then I decided to call up a woman who had told me she would hold her condo until Tuesday so I could decide whether I would rent it for the winter. It was not an ideal place—it was directly on a golf course, had a huge water leak, which means mold, and was permeated with potpourri. But I was desperate and decided to sign the lease. But when I called, she sheepishly told me she had found another renter. Now it was my time to cry. I broke down crying at Laura's house as I was desperately trying to call other real-tors. Sweet little Jimmy, her four-year-old son, gave me a big hug and held my hand as he said a prayer for me. My heart swelled with love and adoration. He truly is a wise and special child.

We raced back to the realtors later that day and looked at three more condos, all of which were no good. With all the stress and the continual exposures of looking at so many houses over the weeks with all their scents and toxic products, my body was losing its steam. I felt I could not go into another house. But for the next few days, I looked at several others until finally I said no more. I had to rest or I would crash so hard I would never get up again. I was completely devastated. I had the stress of living out in the cold and rain, the stress of my mother being upset, the stress of being homeless, the stress of constant chemical exposures from house hunting, and the stress of my failing health. Come on, pile on some more! Instead, God did the opposite.

The woman who had supposedly rented the condo on the golf course called me back and said she felt terrible about the predicament I was in and wanted to help out. She would rent the condo to me if I wanted it. She went on and on about how she and her partner were praying for me and my mother and that they would do anything to help make the condo livable for me. Wow! I was once again blown away by the genuine kindness so many people possess in this world. All those little acts of caring and support really do add up. We arranged for my mother to rent the place for the month of September while she tried to get the potpourri smell out. Then I would be better able to decide if the place would work. Ah, what a relief to have a nice place for my mom to settle down in. She really deserved it!

Once I arrived at Laura's door again, my mother and I decided to stake down my tarp way out in a field in Laura's backyard. It was a stormy and cold day, so my mom insisted I should sleep under the tarp. But the purist (and my stubbornness) in me decided it had cleared up enough that evening that I would be fine sleeping out under the stars. But once I was fast asleep, the rain came down in torrents. I immediately flew out of bed and attempted to salvage my wet bedding. I raced to the back yard (having to make several trips in soaking wet socks) and tried to go back to sleep under the tarp. But since my sleeping bag was soaked, all my clothes were soaked, and my feet were frozen solid, I obviously could not fall asleep. The gale winds, hail, and torrential rains were blowing in the open sides of the tarp, getting me even more wet. This was actually quite a frequent event. I would try to curl up so my body was away from the opening, but many times my feeble tarp would just blow away and leave me out in the rain. This time, after lying there for a while, I decided I would get sick with pneumonia if I attempted to stay out, so I gathered my things, put them all in the garage, and went into Laura's house. The kicker is that after Laura and I had prepared the couch for me to sleep on, I looked out the window and there was my mother's head bobbing along in the rain. It was two in the morning. She had come over because she was worried I was out in the rain in this horrible storm, and of course, didn't think I could handle it myself. Do all mothers feel this way?

For the next several days I attempted to remain in a prone position in Laura's front yard to catch up on some rest. It was getting rather cold already. I was quite the sight when I went to bed at night. I wore my jeans and my snow pants over them, two pairs of wool socks, a cotton shirt, flannel shirt, sweatshirt, sweater, and my hooded sweatshirt over all that. Then I would climb into my 30-below down sleeping bag, though I swear it should be rated a 20-degree bag since I froze all the time. I could not move a muscle until dawn from all the weight. I had my lovely green tarp staked down in Laura's front yard, so I am sure the neighbors just loved the sight of it, or at least wondered what crazy nut would be sleeping out in the snow. Several times I would wake up with the tarp at my nose from the weight of the snow that had fallen during the night. Let me tell you how ready I was for a warm house, TV, and all the wonderful modern conveniences. Luckily, as it got really cold out, I would many times stay in Laura's living room during the afternoons. That was heaven!

At this point my mind was racing with thoughts of where I would go next. By this time my mother and I were too tired to look in other cities up north as we had planned, and we were finding nothing in Estes Park. I knew I had to find something soon because the days were growing colder. In desperation, I called Rebecca, an MCS friend in New Mexico, to ask if she would mind renting me the extra room in her house for the winter while I searched for housing in New Mexico. At this time I was not ready to leave Estes and especially did not want to leave Andy, but I saw no other choice. Fortunately, Rebecca called me back in a few days and said I could rent the room. I couldn't believe it. I really hadn't expected her to say yes, and I felt mixed emotions about this new twist. My only other options were to try to fix up the condo my mom was staying in or rent this other condo I had looked at the previous day, which also had its share of problems and strong scents. Deep down, I knew either one of these condos would be a mistake, but I didn't want to admit it.

Finally, after my mother had used lots of elbow grease, we decided the condo where she was staying would never work for me. With lots of testing and deliberation, I decided to rent the other condo across town. Deep in my gut, I knew this was not the place for me, but I wanted to give it a try. I also just wanted to be out of the snow and cold, sleeping in a warm bed.

For the next few weeks, we desperately tried to fix up the next condo. We hired a maid to help my mother scrub all the walls because she was completely exhausted after all our summer activities. She then proceeded to wash the carpets two times. We took down all the curtains, moved out all the furniture, and proceeded to foil the cupboards. By that time, however, I knew the house was not working out. It still had a strong odor of menthol or something that even a nonsensitive person could smell. We were completely frustrated and stressed, so we stopped all further work until I could go down to New Mexico to test the house there.

To make a long story short, I tried staying at Rebecca's house and loved it. The house was located in a residential area on an Indian pueblo way out near the foothills of the Jemez mountains. The rolling mesas and buttes are absolutely breathtaking. Not to mention, there is no city traffic, noise, or pollution. The pueblo makes money by renting the plots of land that the houses are built on.

The house is gorgeous. Rebecca and her husband totally gutted the place and put in tile floors, white walls, new bathrooms—the list goes on. It is such a light and airy place with lots of windows, just how I like it. The first week I stayed here, my muscles improved, but I developed severe fatigue. I experienced the painful dilemma that week of trying to decide if I would uproot myself from Colorado and make the big move. For years now, I have been drawn to New Mexico. I have always felt at home here, and I knew that eventually I would come to live here. But was it time already? Would I really feel better in this house if I already had fatigue? Was I ready to leave my support system in Colorado and also leave Andy? Those of you who have not suffered from MCS may not realize that trying to find the right place to live is the most crucial and difficult decision to make.

Your health literally hinges on it. I know very few people with MCS who actually have a "safe" place. Many do exactly what I have had to do—live out of their cars, camp, and migrate to wherever they can find a place to sleep that night. Not a very health-promoting lifestyle, but many have no alternative.

After I decided that I definitely wanted to rent the room in Rebecca's house, my mother and I had to go back to Estes Park to pack up my things for the move, which Andy was going to help me make at the beginning of his Christmas vacation. The last two weeks of my mother's stay were hectic beyond belief. Unfortunately, the more time I spent in that condo, the worse my muscles felt. It had gotten to the point that I could do very little around the house again. I could not lift anything above the weight of a coffee mug, I could not walk much, and I was not able to drive again. I would strain muscles by just turning over in bed. It is so amazing how my system can react so strongly to certain houses.

A few days before my mother was to head home, I strained my neck again. We quickly postponed her flight another week while I made lots of calls to local organizations to find someone who could drive me to my medical appointments an hour away in the valley. I finally found someone, and my mother was able to leave at last.

At last I was on my own again. The hardest part of all these years has been the loss of my independence. I have always been fiercely independent—sometimes too much so—and it is a constant struggle for me to strive to be self-sufficient again.

In mid-December, with a U-Haul on the back of my car, Andy and I set off for my new life in New Mexico. He helped to get me situated here in my little abode. We crammed in my computer, TV, VCR, shelves of books, files, clothes, bed bike, heater, air filter, and exercise bike all into this small square room. It looks just like a dorm room. What fond memories it sparks! Andy stayed for the weekend, and then we left one another in the typical airport departure scene.

It has now been a month since I have lived here, and I just absolutely love it here. The desert landscape is so breathtaking that I actually do not miss Colorado one whit. Not to mention that I just love the straight roads. I just put on the cruise, point the car, and off I go. What a sense of freedom again. I go into Santa Fe at least three times a week. Plus my muscles have begun to recover again. I am constantly on my feet—standing and walking is no big deal to me now. I go for walks, ride my exercise bike, and even park as far away from a store as possible so I can walk more. I just love it. Indeed, I am always nervous to state how well I am doing for it is sometimes only temporary, but I hope it continues this time.

In the meantime, I am back to house hunting. It has been unbelievably stressful again, but it helps to know this is where I am supposed to be this time, so God will open the doors I need. Every time I am out driving in my car, I look around me in awe at the vast beauty. I feel so completely at home. My soul is so at peace, I want to jump for joy all the time!

This house that I'm renting a room in was treated for termites five years ago, so I have to research all that was used and run some pesticide tests in the house. Although my muscles have recovered somewhat, I have been becoming increasingly fatigued and "under the weather" here, but it could just be from the gas forced hot air heat, which could be easily removed if I were to buy the place.

Jennifer's update, September 1999: I eventually decided against remaining in that house because it was too hard to drive the long distance into Santa Fe for medical appointments. For the last two years, I've been renting a townhouse in Santa Fe. The summer of 1998 was difficult for me because I developed several lung infections, and one of the medications I had to take may have been responsible for a painful spinal disc problem that kept me in bed for most of the summer. I still have to use a pair of canes to climb stairs if there is no handrail I can use to pull myself up, and my car is specially fitted with hand controls.

Through lots of discipline and hard work, I have slowly been getting better again, however. I am attending a couple of classes at a nearby college and can go out with friends once in a while, but life continues to be a daily struggle.

Alberta

Nursing Instructor

At the school of nursing where I teach, we're doing this marvelous renovation; we're getting all these new cubicles, carpeting, and new furniture. I noticed that after being there for a couple of hours, the smell of the treatment or chemicals in the carpet gives me a headache. I have to be very careful being around the new furniture because the particleboard it's often made of and all those things they treat furniture with in storage will give me a headache and an upset stomach. My head hurts, my eyes start watering and sometimes itch. I get stuff running down the back of my throat and I feel like it's going to cut my breath off.

I used to think it was only certain oil-based paints that bothered me, but any kind of paints will give me a reaction now. If I walk into a building and they're painting, I back out, get outside, because it always makes me sick. I have a tendency to migraines and what can happen is I'll go into one of these situations with carpet or paint or whatever, and I get what I call an allergic headache and if I stay in that situation too long, I end up with a horrible migraine. I'm a very busy person, I'm a mother, a wife, I'm very active in my church, and I teach, and it takes a fair amount of my energy. I just don't have time for that.

I sing in a church choir, and I run into problems with fragrances there. People have a tendency to put on cologne when they get out of the shower, and you have all the different kinds of lotions and the powders and the perfumes that go together, so why not put them all on at one time? When you get 30 or 40 women in a closed space like a choir pew, all these odors start coming after you. Then someone pulls out the lotion they forgot to put on when they left home, and after they put it on their hands, everyone else wants some. By that time my eyes are running, there's stuff running down the back of my throat, my head feels this big, and I'm glad I'm not a soloist because I wouldn't be able to get a note out.

We have asked that people not do that, but they don't think. They put on cologne and they like the scent and want to put on more, but it's just overpowering to me. I've had to move away from people in a crowd because they have a strong lotion on or hairspray or something like that that bothers me.

Moth balls are horrible.[1] You walk by a little old lady that has had her stuff in moth balls for the summer, and it's nauseating. It's those simple, everyday things that can make you very ill that you have to figure out how much of it you can inhale. You try and get away from it as quickly as possible. It's really a problem if you're in a crowd of people where you have nowhere else to go and it gets warmer in the room and you have more of these scents coming after you.

Going to a hotel can bother me because of whatever it is that they use on the sheets. By the time I wake up in the mornings, I have this dull, achy, heavy feeling behind my eyes. If it got severe enough, it would make you not want to go anywhere because you don't feel well so much of the time. I have a relatively high tolerance for discomfort, however. I'm just too ornery to give into things. I don't necessarily let it stop me. I try to figure out ways to get around it.

In a lot of nursing facilities like the school of nursing where I teach, they clean things frequently. Nothing that would be harmful probably to the normal, average person because of OSHA regulations. You do have to be careful what you have around your patients because your patients might have certain sensitivities. In my case, I can have a reaction if I walk into a room and they're using something that has maybe a chlorine base to it and smells like typical household bleach. That might bother me, so I leave the room. My basic management of these difficulties is avoidance because that's all that I can do. These are things that exist in the everyday world. Most people don't think about it because they don't have the problem. So you can't say, "Please don't use that cleaner when I'm up here" because I'm the only one it bothers, so I use avoidance.

Basically the way I manage these chemicals that bother me is by avoidance, which keeps me healthy enough so that I can work every day. You need to know what you need to get away from and how fast you need to get away from it in order to not become ill.

---

[1] One leading brand of moth balls states on the box: "HAZARD TO HUMANS AND DOMESTIC ANIMALS. May be fatal if inhaled." It's not clear how one opens the box without inhaling vapors.

Ambrose

Vietnam Veteran

I'm a member of the Navajo tribe, and I live in New Mexico. My health was excellent when I was growing up, and I was very athletic when I was in high school. I played football, ran cross-country, and also was a swimmer.

Right after I graduated from high school in 1965, I was drafted into the army and sent to Vietnam. After my advanced infantry training, I was supposed to be a mechanic, but instead they assigned me to drive a jeep. My job was to drive officers. Our unit was supposed to be supporting the combat troops, and I often drove officers through the area between Saigon and the demilitarized zone.

One time we were part of a convoy escort that went way up north. After we left the company there, the executive officer I was driving wanted to go back to Saigon. We drove back by ourselves, with no escort. We were driving through the jungle when we noticed a crop duster plane flying low and spraying the rice paddies and orchards. I had seen crop duster planes spraying wheat and alfalfa fields while I was working in Idaho one summer when I was in high school, and that day in Vietnam I thought they were just spraying some insecticide that wouldn't hurt people. As we were driving through the Vietnamese jungle, it happened to be a very hot and humid day. The cool mist from the crop duster felt great as we drove through it, so I stopped the jeep, stripped off all my clothes, and started running through the mist to cool off. We were out in the jungle with no one around, so I didn't worry about anyone seeing me. After a while, I got dressed and we drove back to Saigon.

A month later we were driving through the same region, and we noticed that all the vegetation was turning brown, but we didn't think anything about it. No one had told us anything about Agent Orange being used in Vietnam. We even spent several days camped in the middle of that area.

Before my tour of duty was over in Vietnam, I got wounded in the head by shrapnel and was paralyzed, so they sent me to Japan for treatment. After a month, the paralysis disappeared, and they shipped me back to Vietnam, where I spent two more months before my term in the army was over.

I got out of the army in 1968, and about three or four years later I began getting funny feelings in my legs. My feet were always cold, even in the summer. I had no feeling in my skin below my waist. If a fly landed on my leg, I could see it walking around but I couldn't feel it.

When I began to hear about the use of Agent Orange in Vietnam, I went to my tribal office to see a VA officer who was supposed to help veterans who thought they had been injured by Agent Orange in Vietnam. He had a picture hanging on the wall that showed a crop duster being used in the jungles of Vietnam, and I told him that that was the kind of plane that had sprayed me. The VA officer explained to me what was in Agent Orange, and I told him about how I had bathed in the mist from that crop duster.

I never succeeded in getting any compensation from the government because they said any symptoms I had from Agent Orange should have shown up within a year. After I risked my life and was wounded in Vietnam and exposed to Agent Orange there, I think the government should be helping me now that I'm in bad health.

Three years ago, when I was 53, I had to give up working as a mechanic because I couldn't handle tools anymore. I kept dropping them. When I was around gasoline and oil, my skin would break out in rashes and then peel. Now I have periods where I'll be driving down the road, and I will find myself in a different place and have no idea how I got there. The doctors say they don't know what is wrong with me. I get tired a lot now and can't walk more than a few minutes before I get short of breath. I can't do any heavy work or lift heavy things.

I can't stand to be around cleaning products now. When my wife uses different things to clean the house, I have to go outside. I also can't stand perfume; when people wear a lot of it, I get a headache.

My wife and I live on a mesa with our sheep, cattle, and horses. Our nearest neighbor is seven miles away, so there's not much pollution out here.

Editor's note: Native Americans have less of an enzyme called alcohol dehydrogenase, which is used to detoxify alcohol and is also used in other detox pathways. The New Mexico prevalence study of MCS found that an unusually high percentage of Native Americans reported that they are chemically sensitive. In response to the question: "Compared to other people, do you consider yourself allergic or unusually sensitive to everyday chemicals like those in household cleaning products, paints, perfumes, detergents, insect sprays, and things like that?" 27% of the Native Americans contacted replied yes. The percentage of the overall population responding yes was 16%. (The Anglo and Hispanic groups were statistically indistinguishable from each other.)

Karen

Legal Secretary

I got chemical sensitivity in 1990 when I was working as a secretary for a law firm in a high-rise office building. In my case, carpet removal and installation contributed to my getting sick. They used a solvent to take out an old carpet, and a day or two later they glued down a carpet on my lunch hour, so I was exposed to lots of volatile organic compounds.

When I first began to get ill, my symptoms—nasal irritation, burning sinuses, and a sore throat—were quite tolerable. I also had some respiratory effects, but nothing I worried too much about. I had these symptoms for months, and they seemed to be getting slightly worse. Finally, I noticed that my ear was feeling as though it was swelling closed, and it felt like I had tunnel vision in my right eye. Within ten minutes of getting to work, my face would get numb on one side, so that was a whole new level of concern for me. I still had not really attributed my symptoms to my work-place, however. I made that connection when I realized that I was always reminding myself to bring my decongestants to work with me because that's where I was needing them. So I was just medicating to allow myself to tolerate those symptoms, but they continued to get progressively worse.

The point at which I began to feel like I could no longer tolerate my symptoms was when they started to affect my ability to function. I was having problems with slurred speech, and I was having problems thinking. I couldn't find words, and I couldn't keep my train of thought. It was so bad that there was one day when I was trying to write a letter and I couldn't think what day it was. I just worked and worked to think what day it was, and finally I found myself looking out the window, trying to narrow it down to a winter month or a summer month. At that point I felt like I was in trouble.

I had been going from doctor to doctor to try to get relief for these symptoms. I figured it was allergy or whatever, and I finally was diagnosed with migraine and tried to take migraine treatments. And because of all these neurological problems that I was having with my ability to speak or write, I finally couldn't proofread at all because I couldn't hold a sentence in my mind long enough to tell whether it made sense.

Then I began to have reactions when it felt like my nervous system was strobing or tremoring constantly for weeks at a time. I had one final event when that strobing amplified and I had to leave work. I fully expected to be back at work the next day, but the amplification didn't resolve, and it didn't resolve for seven days, so I never made it back to work. I had by then developed sensitivity to chemicals. I went from sick building syndrome to multiple chemical sensitivity.[1] During the last few months, I had noticed that there were things at home that would trigger my symptoms. I couldn't have a newspaper in the house. When I used a chlorine cleanser in my sink, I got very sick from that. I got very sick washing the windows one day. All my reactions were neurological; I would feel like my brain wasn't working right.

I really would have expected to be the last person in my office to get chemical sensitivity because I worked only three days a week and I was so much more active than the other people that I was working with. I had grown up taking dancing lessons from the time I was four years old right through my teenage years. I had been involved in sports, particularly on baseball teams and volleyball teams, and my husband and I had done a lot of backpacking and mountain climbing. So I was just surprised that I could develop MCS. Everyone is at risk. It's simply not a predictable thing, who will get chemical sensitivity.

When I was not able to return to work, I found it hard to believe a building could make you that sick and you could lose your ability to work because of exposures in your building. So I decided other people needed to know about this, and I called the newspaper. They were interested in the issue of sick building syndrome because our state was beginning to give more attention to air quality issues in constructing buildings, and the newspaper thought the stories tied together well. Well, when the story came out, I was very surprised to have the phone ring off the hook, and 35 of the calls that came in during a four-day period were from people who had stories like mine. I took their phone numbers and their names and thought that we might meet or have some way to communicate with one another. Some of us did meet, and we decided that this wasn't OK to lose your livelihood from exposure to things in your workplace, so we wanted to raise awareness about chemical sensitivity. I continued to collect names, and I was asked to join an indoor air quality group where I worked with folks in agencies and in county health departments. Because of my work on this committee, people

---

[1] A distinction sometimes made between sick building syndrome and MCS is that the former term should be reserved for cases of symptoms arising from chemical or mold exposures in the workplace that disappear when one leaves the environment. A certain percentage of the people who develop sick building syndrome because of workplace exposures then go on to develop MCS, reacting to low-level chemical exposures even when they are away from their workplace.

would refer others with MCS to me. I became an advocate, providing resources and referrals and information because I got into the research and felt as though education was really important to people with chemical sensitivity. During the past nine years, I've assembled a database containing over 800 names of people in the state of Washington who have developed multiple chemical sensitivity.

In talking to other people who have developed chemical sensitivity, I have heard story after story of people who lose their jobs, who lose their health insurance, who take the fast track to poverty at the same time that they're really very ill. I've been more fortunate than most people with MCS because my employer was very caring and concerned. My husband was very supportive and was willing to make the changes that he needed to make in his lifestyle to accommodate me. I have health insurance through his employer. I have a supportive family and friends, and they did not blink an eye at having to reduce their use of colognes, perfumes, hair spray, and laundry products around me. They didn't want to let my chemical sensitivity interfere with our friendship, so they just made those changes for me. I'm also fortunate to live in a home that was built in 1923, and my neighbors are not too close, so I have a safe place to live. Another way in which I think I'm very lucky is that over these ten years since I began to get sick, after a period of three very sick years, I have begun to get better. While I'm still extremely sensitive to chemicals and can get a migraine from smelling wood smoke for a short period of time and perfume provokes my symptoms almost immediately, I feel really pretty well when I am not exposed to things to which I react. Unfortunately, however, my chemical sensitivity still prevents me from returning to work.

My friends sometimes forget I have chemical sensitivity because it doesn't show on my face. I don't look ill; people tell me all the time that I look well. But I could be having a migraine at the moment they're thinking those things, and they would never know unless I told them. Obviously there are symptoms of chemical sensitivity like rash or eczema that you would see, but many, if not most, of the symptoms don't appear at all. That's sometimes difficult for us, having an invisible disability.

I have not been faced with a lot of the things other people have been faced with. I cannot imagine, for instance, losing your job, losing your health insurance, being a single parent with a mortgage. What are those people doing? So I've wanted to help them find whatever support network is available. I've learned that getting Social Security benefits might be a two-year process, and I try to help people find a lawyer who will assist them if they have a claim. Lawyers are not inclined, however, to take cases like these. That's a big difficulty.

Finding housing is another enormous difficulty, especially with no income. There is a three-year waiting list for subsidized housing in this area. Most newer housing has been constructed using cheaper building materials like particleboard that tend to outgas more formaldehyde and other volatile organic compounds. There is frequent use of pesticides, and some neighbors smoke. As a result, these chemically sensitive people who fall into that track have a life full of despair and desperation, and they find themselves suddenly struggling for their survival.

MCS is just so hard on relationships for many reasons. People with chemical sensitivity have to control their reactions by controlling their environment. And it's one thing to control your own environment, but there are many occasions when you have to control somebody else's environment, and that may be very difficult to do in a household. Many people with chemical sensitivity have teenage kids who are using all sorts of hair products or other scented cosmetics. And some women have husbands whose hobby is woodworking, and they are sawing and staining and doing all these things that affect a person with chemical sensitivity. It really makes for some very difficult times when one's partner or children suddenly and unexpectedly have to make all of these changes. I know my own husband had spent many years with me being very well and healthy, and then after my injury I was on the couch most of the time for a couple of years. When I had migraine headaches, he most certainly couldn't count on having dinner on the table. The level of cleanliness in my house also went way down because I just couldn't do all the things I had done in the past. And I had to give up doing chores that I had done for years and years. I just didn't have the strength to do them in some cases or they involved an exposure that just was not going to be an option for me anymore.

It isn't just the chemically sensitive person who becomes limited by the disability, the entire family becomes limited. There are lots of places that just aren't going to be in the picture for a couple. They may have gone to movies regularly or gone out to dinner or dancing or any of those things, and all of a sudden, bam, not an option anymore. And that spouse may have really just loved socializing. That's been a difficult one for me because I'm really a social bug and have always been. I have lots of friends, and I like dinner parties, and I'm just so lucky that my family and friends have been willing to be fragrance-free because otherwise I simply would not have those wonderful things in my life anymore.

Over the last eight or nine years, I've talked to lots of people whose marriages didn't survive the spouse having chemical sensitivity. The well spouse just was not able to make the change. Sometimes a spouse is unwilling to give up wearing perfume or aftershave. A woman I talked to the other day said her husband cannot stand having her wear a mask in the car. He is very embarrassed when she does that. And a lot of spouses can't tolerate having a sick person around all the time and someone who has to be so very controlling or who might be so very irritable or maybe so dull and fatigued or whatever. It just is quite a burden on a household.

Most people have a hard time believing that people can be this sensitive to chemicals, and I can see why they do. When I talk to people and they tell me they are reacting to such and such, I sometimes think, "Could they really be reacting to that?" I have to remind myself that I used to be that sensitive. At one point I was down to two foods that I could tolerate. Now all these years later I tolerate all foods, and that symptom has pretty much disappeared. So I understand disbelief, but I also know how important it is for outsiders to try to understand chemical sensitivity. Losing your health, losing your income, losing your health insurance, becoming isolated, losing friends, those are all very difficult things for anyone to have happen in their lives. But on top of these problems, it is hard to have to try to convince people that you are ill and that chemicals are making you sick. In many cases you encounter skepticism or attacks on your credibility, and some people try to suggest that you are psychologically unstable. Facing those attitudes creates another enormous burden on a chemically sensitive person.

Alison

Freelance Editor

My daughters and I are among the lucky people who developed MCS and have recovered sufficiently to be able to live our lives without excessive restrictions. There was no obvious exposure that made us all sick, and the reasons why each of us developed MCS years apart are not at all clear. My surveys and extensive personal contacts with people with MCS, however, have indicated that it is highly likely that there is a genetic predisposition to develop MCS.[1]

One of my first major exposures to chemicals occurred when I was about 11 years old. For a year or two, I moved out of the bedroom I was sharing with an older sister and set up a bedroom of sorts in the corner of our unfinished basement. I slept right next to our coal furnace. It was my job to remove the clinkers from the furnace, so every night before I went to bed, I would lift them out into the bucket sitting by the furnace door, where they cooled all night, giving off acrid fumes as I slept only a few feet away. Thinking back, I remember that this was a period when I had constant colds and bronchitis.

Years later when I got married and moved to Maine, I once again had some extensive exposures to chemicals. I fell in love with New England antiques and would go to antique shops and auctions to buy bargains that could be refinished. Over the next half-dozen years, I refinished at least 18 pieces of furniture. I removed coat after coat of paint with paint stripper, which is a fairly potent solvent. The labels on the cans said to use with adequate ventilation, so I would open a window a few inches when I had to strip a piece of furniture inside because it was too cold outside for the paint stripper to work.

My husband and I also did all the redecorating work that we were capable of doing ourselves when we bought an older house. We stripped ugly

---

[1] Robert Haley, M.D., Department of Internal Medicine, University of Texas Southwestern Medical Center, Dallas, Texas, has published results of a study showing that Gulf War vets who developed Gulf War syndrome have lower levels of an enzyme that hydrolyzes several organophosphates. (Many leading pesticides are organophosphates and act as neurotoxins.) Another group of researchers has published a study titled: "Mice lacking serum paraoxonase are susceptible to organophosphate toxicity and atherosclerosis," *Nature*, July 16 1998, pp. 284-87.

varnished woodwork all through the house, we painted walls, kitchen cabinets, and woodwork, and we varnished floors. For months we were breathing toxic fumes.

Another thing that fascinated me in New England was the beautiful hand-braided rugs I saw in older homes. I decided to braid one, so for many months I often had pots of dye and wool simmering on the kitchen stove. I can still remember how strong the hot dye smelled. But at that time in my life, I did not notice any effect from all these exposures.

One other exposure that I've sometimes wondered about occurred when my husband had a sabbatical and we spent a semester in Paris in 1971. There was a gas hot water heater in the kitchen of our apartment, and I have a clear memory of walking into the kitchen every morning and flinging open those tall French windows to let in as much fresh air as possible because I could smell gas fumes.

The next winter I happened to have a fairly heavy exposure to cigarette smoke because I was on a committee that met in my kitchen a couple of nights a week and one member was a heavy smoker. That, of course, was back in the days when you didn't dare ask anyone not to smoke in your house.

That March I had pneumonia, and about three weeks after I recovered from that, I began having migraine headaches for the first time in my life (I was 33). After I had had several of these migraines, I made the connection that they always started early the next morning after I had been exposed to heavy cigarette smoke the previous evening. I can still remember lying in bed feeling sick as a dog with a throbbing headache and frequent vomiting or dry-heaving, while my second daughter, who was just two and a half, wandered around the bedroom trying to get me to read her a story or play with her. One interesting aspect of these migraines was the ten-hour delay between exposure and reaction.

After a few months, I also began to see that exposure to heavy diesel fumes could trigger a migraine, sometimes with only a few hours delay. Then the following October I began to have so much joint pain when I got out of bed in the morning that it was difficult for me to pull my nightgown off over my head. When I walked downstairs, my left knee hurt, and it was painful to turn my left hand into position to play the violin. These symptoms were totally new to me; I had never experienced any joint problems previously. I had tests for rheumatoid arthritis, but they were negative. Finally, it crossed my mind to wonder if my joint pains could be related to our oil furnace going on at the beginning of October.

The connection between my bout with pneumonia and the development of chemical sensitivity three weeks later seems quite apparent, and I always used to say that pneumonia was the precipitating factor for me. As I've thought over the chronology in preparation for writing this story, however, I now see that it is quite likely that the heavy exposure to cigarette smoke precipitated the pneumonia and the double assault on my system from the smoke and the pneumonia produced the chemical sensitivity. And of course my earlier exposures to coal smoke, solvents, and dyes could certainly have "primed" me to develop MCS. As in so many MCS cases that I have heard about, it is often difficult to determine which factors were paramount in the development of chemical sensitivity.

Back in 1972, when I first experienced clear symptoms of chemical sensitivity, there was virtually no information available about the subject. At last I did find out about the work of Theron Randolph, M.D., the "father" of the field of environmental medicine, who lost his position on the faculty of Northwestern University Medical School after he began talking about chemical sensitivity. I found Dr. Randolph's book *Human Ecology* to be very helpful, and it was a great relief to find a physician at last who understood the unusual condition I had developed.

In the early 1970s, only a handful of doctors in the entire country were familiar with chemical sensitivity and were attempting to treat MCS patients. There was not the present variety of therapies to choose among. My only option was avoidance of chemical exposures and problem foods, which still seems to be the most effective way of recovering some degree of tolerance. I started preparing almost all of my food, including bread, from scratch, and I started using as many organic ingredients as were available. In addition, I drank spring water to avoid chlorine. I decided to go on a rotary diet, which I followed for about a year. On this diet one eats a given food like wheat or carrots only once in four days.[2] The diet did help me see, for example, that coffee or chocolate, anything with caffeine in it, could cause a migraine and onions could make me feel unwell. When I eliminated those foods and drinks from my diet and avoided cigarette smoke for a year, my migraines disappeared. I only had migraines for two years and suspect that in many people migraines can be traced to sensitivity to various chemicals or foods.

My husband and I also decided to make major changes in our home to reduce my exposures to chemicals. The main thing we did was to replace our oil furnace with an electric boiler to heat the water that went through our hot-water radiators. Heating with electricity in Maine is expensive, but good health is worth it. Not only did my joint pains disappear, but from that

---

[2] For an excellent description of the rotary diet, see Doris Rapp, M.D., *Is This Your Child: Discovering and Treating Unrecognized Allergies in Children and Adults* (New York: William Morrow, 1991), pp. 457-80.

point forward, my husband and I almost never had a cold. We also went through the usual MCS routine of changing over to less-toxic cleaning products, laundry detergent, and personal care products.

I was fortunate because at this time I was working at home as a violin and piano teacher. Younger students didn't wear perfume, and I could ask older ones not to wear perfume or use a particular shampoo that might be bothering me. Eventually, I decided to switch from giving music lessons to doing freelance editing, which gave me more flexibility to choose my own working hours. Had I been working outside my home, it would have been far harder to recover my health.

Within the six-month period after I started the rotary diet, removed the oil furnace, and generally eliminated other sources of toxins from our house, my health went back to normal. No more arthritis, no more migraines, and I was no longer hypersensitive to perfume. We were even able to conceive another child, which we had unsuccessfully been trying to do for four years.[3]

I am grateful that my husband never questioned the reality of my chemical sensitivity. Like me, he was ready to look for cause and effect. When I had a reaction, the question was what had I eaten or been exposed to, not whether I had had a stressful day. For a year he did all of the grocery shopping so that I could avoid the cigarette smoke that used to fill so many stores.

I'm in good physical shape now. Last summer I had just turned 60, and as usual I did a lot of high-altitude hiking in the Rocky Mountains. On one occasion I hiked 19 miles round trip with a 5000-foot altitude gain. Before we made the switch to an electric boiler for our heating system, I was having knee pain going down a flight of stairs—descending steep mountainsides would have been out of the question with that level of joint pain. My health in general is excellent; I never have the flu and rarely have a cold. Apart from the occasional physical, I only see my doctor once every couple of years.

There is hardly any physical symptom I experience that I can't trace to an exposure to a chemical or food. Once my left heel suddenly became so sore I could hardly walk on it. I realized at once that I was probably reacting to the heavy pine terpenes from a couple dozen freshly cut pine boards I had picked up at a local mill and put in the basement so I could make some shelves. Out went the boards, and the heel returned to normal.

On another occasion I photocopied a lot of pages that ended up with large areas of black ink on them. When I was sorting them out at home a

---

[3] My survey results suggest that chemical sensitivity may reduce fertility. See my article titled "Reproduction" listed in Appendix 2.

half hour later, my heart began to feel as if it were flopping around in my chest and my heartbeat became very irregular. Anytime I spend too long using a photocopier, I notice irregularities in my heartbeat. Exposures to moth balls will produce this same irregular heart beat.

I have recovered from MCS to a large extent, but I still have to be aware of my underlying chemical sensitivity. It places no great restrictions on my life at this point, however. I go where I want and do what I want in general. It's rare for me to think that I can't do something I want to do because I have MCS.

When our daughters began, one at a time, to develop MCS a few years later, we understood the condition. When one daughter started getting dizzy at her summer job at a drive-in restaurant, we asked the right questions and found out she was frequently waiting for orders near a large gas range. When she went to Stanford and after a few months suddenly started having vomiting episodes, we asked about exposures and learned that a new carpet had just been installed in the common room on the first floor of her dorm. By walking quickly through the first floor area to get to her room on an upper floor, she was able to eliminate the vomiting. It came back three months later, however, when they carpeted the hall outside her room. By keeping her door shut and holding her breath as she walked down the hall, she was able to minimize the problem.

Another daughter would get extremely weak just from sitting in a bathtub of our regular municipal water, which contained a high level of chlorine. We eventually had a whole-house filter put on our water system. The car heater gives a couple of us headaches, so we never use the heater, even in the frigid Maine winters.

When one daughter had her first migraine when she was about 12 years old, I immediately asked what she had eaten that morning and the culprit turned out to be a chocolate-covered cherry. She didn't have another migraine until a few months later, when she happened to eat a cherry tart that was dyed with the same bright red dye.

We weathered some very difficult times when we were raising our daughters. When one of our daughters was in the eighth grade, there was period of a few months when she would wake up every morning totally exhausted and unable to get out of bed until the middle of the morning. Her condition worsened until there was a period of six weeks when she couldn't attend school at all. In her case, we learned by taking her to Colorado, a much drier climate, for a few months that mold was a huge component of her problems. (Mold gives off toxic chemicals.) By the beginning of high school, she was well again, and a couple of years ago she ran a marathon.

It wasn't easy raising daughters with MCS, but we kept them healthy enough to be starters on their basketball teams, and they eventually got degrees from Harvard and Yale law schools. One reason they have been able to stay healthy is that they always look for apartments with hardwood floors, electric stoves, and hot-water radiators. Although my daughters have been leading "normal" lives for some time, they can only maintain this status by rigorously avoiding exposures that they know have caused problems for them. For example, my oldest daughter has been working as a lawyer for a large New York law firm for six years. Her job involves long hours and a fair amount of international travel. She has been very fortunate that the indoor air quality at her law firm is excellent. Nevertheless, exposure to perfumes on colleagues or clients, as well as air fresheners in taxis or public restrooms, can set off headaches, dizziness, and cognitive symptoms that interfere with her capability to work for hours at a time. Any exposure to new carpeting, paints, solvents, or offgassing construction products produces a similar effect.

Avoidance of exposures has been the one thing that has worked to return all of us to health. We each still have some chemical sensitivities, but we can work around them well enough to live full and productive lives.

Eli

Teacher

I had a mild case of MCS for a decade. For example, the furnace that was built into the floor of the house I rented leaked gas, and sometimes those fumes would give me a headache or flu-like symptoms. I had no idea, however, that any illness could change my life. I taught mathematics in elementary and middle schools and also did some tutoring. I enjoyed swimming, hiking, and gardening for recreation and could still do what I wanted to do at this stage of my life.

On February 17, 1994, I was living in an apartment in a house at the edge of open space where my landlord had agreed not to use pesticide. Before that day, I had never had any breathing problems. On that day, however, I awoke from a nap, a much longer one than usual, feeling drugged. When I opened the door of my bedroom, where the windows had been closed, there was an overwhelming pesticide smell inundating the rest of the apartment, wafting in through the open windows. I felt as if I couldn't breathe, so I immediately went back into my bedroom, shut the door, and turned my carbon air filter on to the maximum setting. After a while, I could breathe a little better, but within a few days it became obvious that exposures to chemicals like rubbing alcohol could now cause breathing problems. When I went to my doctor of traditional Chinese medicine, whom I had not yet told about the incident, she listened to my pulses and exclaimed, "What happened to your lungs?"

After this pesticide incident, my energy level dropped. The pesticide exposure also seemed to weaken my immune system; I had never had an intestinal parasite problem before, but within a month of the pesticiding, I began having them persistently.

My landlord was evasive when I told him about my breathing problems and asked if his gardener had applied pesticide that day. The most he would say was, "We could look at this as something that will not happen again." Months later he wanted to use a preservative to prolong the life of his roof. When I researched the preservative and told him my lungs probably wouldn't tolerate it, he grudgingly agreed to hold off on using it. I naively saw this as a change of heart for the good.

Two weeks later, on November 29, 1994, when I came home from work and walked into my apartment, I immediately became nauseous and felt like

I was going to pass out. I ran out of the apartment to get some fresh air, and then a few minutes later, I tried going back in. The same feeling hit me again. To this day, I don't know what had been done to my apartment, but there was a thick, greasy residue on some of my belongings. I attempted to have the apartment detoxified, but I was never able to live there again. In addition, I could no longer use any of my possessions that had been in the apartment. Even after I washed my clothes repeatedly, I would get rashes when I wore them. Handling my papers or books made my hands tingle.

After this second exposure to what I assume was pesticide, I became even more sensitive to such chemicals. If I happened to run into a pesticide somewhere, then that night when I was in bed, my limbs would twitch and jerk and sometimes grow numb. I would occasionally have massive breathing problems.

Fortunately, I was able to find shelter with a family whose children I had tutored. I became increasingly aware, however, that my new neighborhood had frequent and large-scale insecticide applications. Between the second exposure in my apartment and the spraying in my new neighborhood, my immune system was weakened still further. I caught the flu, and at first gave it no thought because I had previously always gotten over the flu after a few days or perhaps a week or two. This flu persisted, however, and it began to seem as if I was picking up viruses the way a magnet picks up iron shavings. As a result, on the days I managed to make it to school, I could only teach a couple of classes. I began to fear that I could not continue to teach. I had always prided myself on getting my students to develop their creativity and their reasoning powers. Now more than ever, I put heart and soul into what work I could still do. Though there was much pain in my life as my health deteriorated, there was also great light in the classroom. I will never forget the discoveries my students made during those difficult days.

While I was teaching a reduced load in 1996-97, I still had hopes that I could go back to work when my health improved. I got substantially worse, however, and was diagnosed with chronic fatigue syndrome, as well as MCS. I had to give up on my career and fight for my life.

I would never have believed there would be a time when I couldn't work. An elderly relative who was concerned about my condition had asked if I might have difficulty making a living in the future. I replied: "I plan to teach to the day I die. But if I can't teach, then I'll tutor. If I can't tutor, then I guess I'll do something from home with a computer. I'll always be able to do something." My relative was reassured, willed a small fortune to someone else, and died.

For 25 years, I taught, tutored, and did volunteer work, always helping others. Now I must have others help me with shopping, household chores, and personal care. Preparing meals and doing the dishes each day seems as much of a challenge as climbing Mt. Everest. If I get fed up because I can do so little and I try to do something more around the house, I sweat and tremble, lose my balance, and nearly collapse from exhaustion. My feet become so painful that I can't stand on them or even sit with my shoes on.

Before I was exposed to the pesticides, I had a good back. Now problems in my back and neck prevent me from driving. I have to lie down in the car while someone else drives. Even this motion aggravates my back problems.

Even without my back problems, my days of driving would have been numbered because I began to lose the strength in my wrists, shoulders, and feet. Neurological problems like visual migraines and balance problems plague me. If I have a minor chemical exposure during the day, then at night I feel like my head is swirling around when my eyes are closed.

Taking a shower has become a huge effort. It's a big strain on my back to get my legs over the side of the tub. I manage this by placing a plastic chair so that two of its legs are inside the tub. Then I can sit down on the chair and swing my legs into the tub. Even moving the chair takes a lot of energy and puts stress on my back. If I shower in the morning, I collapse for the rest of the day. If I shower in the evening, I can't sleep and my arms turn numb. In either case, I experience visual and neurological distortions.

I thought I had my parasitic infections under control after five years, but I seem to have another one now. That worries me because there is no longer an efficacious antiparasitic medication, prescription or natural, that doesn't make me sick, nor can I take antibiotics. Organisms that are held in check by a normal immune system have become a big problem for me.

At least when the chips are down, you find out what's in someone's heart. My family support has been gratifying, and this is the third year that my school has given me a paid leave. Former students and parents have reached out to help. A friend does my laundry. My place of worship lets me attend a study group by phone and announces my need for a safe place to live. Neighbors agree to use less toxic methods of pest control or at least warn me when they are going to use pesticides. Friends shelter me when I need to leave home for a while. Caregivers for the disabled take a personal interest in me. Landlords rent to me, though it would be easier to rent to someone else. Generous and kind-hearted healing professionals have given far above and beyond what might be reasonably expected from them. All these kind acts help keep me going through an unbelievably difficult time of my life.

At this point I am so sensitive to pesticides that even with the windows closed and sealed shut, the door closed and covered with a sheet, and many air filters running, when a neighbor sprays certain pesticides my skin burns, I lose my balance, and I have difficulty breathing. I also have visual problems and increased numbness, pain, twitching, and jerking. My eyes become so sensitive to light that I'm bothered even by having a light bulb on. And all these reactions result from a tiny fraction of the dose of pesticide I would get if I were outside.

The reason I keep going on from day to day is not because life is bearable but because I hope to help educate other people about the dangers of pesticide use and the chemical way of life. By telling them about less toxic and more sustainable alternative ways to live, I hope they can avoid what has happened to me. As one torture victim wrote to another one who later became an Amnesty International leader: "Do everything you can to survive. There will be other human beings in the same situation as we are. Let your voice be heard."

Afterword

These personal histories raise many important questions and issues not only for individuals with multiple chemical sensitivity, but also for society at large.

1. Is MCS simply a condition that affects just a few unfortunate people or are growing numbers of people at risk in our society, as we rapidly expand our use of toxic chemicals in careless and unnecessary ways? Appendix 5 contains brief summaries of several articles from scientific journals or other media sources detailing the harmful effects of various widely used chemicals.

The personal histories in this collection come from people who have made the connection between their illness and chemical exposures, but many of them described a long period in which they were feeling worse and worse and couldn't figure out why their health was declining. There are undoubtedly many people in the general public who could eliminate some troublesome symptoms like migraine headaches or arthritic pain by considering possible connections to chemicals or foods.

In addition, it is evident from these personal histories that most people with MCS have periods when they simply can't think clearly or are especially clumsy, even walking into door jambs. If chemical exposures are affecting others in the workforce to a lesser degree, businesses may be losing significant amounts of worker efficiency because of poor indoor air quality. In my third MCS video, Dr. Gerald Ross cites a study of a "sick building" that shows that even those workers who claimed they were experiencing no symptoms showed a decreased level of mental functioning.

2. The rapid growth in the number of people who have asthma has become a pressing public health issue. In many newspaper and magazine articles, reporters state that no one knows why there is such a sudden surge in the number of people developing asthma. Surely it is suggestive that so many people providing their personal histories for this collection report a sudden onset of asthma that was clearly triggered by an exposure to toxic chemicals.

Understanding what Dr. William Meggs calls chemical irritant asthma is crucial in making public health decisions affecting asthmatics. Recent research, for example, has shown that many asthmatics are very allergic to cockroach droppings. This research has unfortunately led to calls for increased spraying of pesticides to kill the cockroaches. Asthmatics who are sensitive to chemicals are likely to be made worse by this pesticide spraying, however, and would benefit more from the use of baits or boric acid to combat the cockroaches. The potential for pesticides and other chemicals to trigger asthma attacks is usually not recognized by the people making decisions about the use of these chemicals. Consider the headline of a *New York Times* article on March 5, 1999, which stated: "Asthma Is Found in 38% of Children in City Shelters. Rate Is Six Times the Average for Children." Shelters are likely to be places where there is very frequent use of pesticides, disinfectants, and air fresheners, which may well be producing or exacerbating asthma in some children.

3. Gulf War syndrome now affects over 100,000 Gulf War vets. This development has not only produced severe health problems for the ill vets, it has also raised troubling ethical questions for society in general. The men and women who went to the Gulf War risked their lives to serve their country and deserve far better treatment than they are receiving. It is crucial that researchers look more closely at the relationship of Gulf War syndrome to MCS for several reasons.

If there is a strong connection between the two syndromes, then avoidance of chemical exposures may be one of the only therapies that can help the ill vets at this point. Medications must be used with caution for the chemically sensitive because they often simply exacerbate the situation. When PBS did a special on Gulf War syndrome in 1998, one of the vets featured who looked very sick actually had a job pumping gas. That kind of intense exposure to a toxic substance may only be making him sicker.

If a large percentage of the ill Gulf War vets do in fact now have MCS, as some surveys have suggested, then it is essential for the government to realize that MCS is not simply an inconvenient condition in which people are bothered by perfume and diesel fumes. As the personal histories in this collection illustrate, MCS is a condition that can all too quickly make it impossible for a person to work because most workplaces involve exposure to chemicals that those with MCS simply cannot tolerate without becoming very sick. Thus service in the Gulf War may have made many vets almost unemployable, and they are left struggling to find a way to provide for their families. Vets who find themselves in this devastating situation are entitled to disability payments from the government for the period during which their disability persists. This will be a large expense for the government, but we have a clear moral duty to see that these vets who faced danger on behalf of their country receive the benefits they deserve.

In the November 19, 1998, hearing of the Presidential Special Oversight Board investigating Gulf War syndrome, which is chaired by Senator Warren B. Rudman (New Hampshire), Major Denise Nichols, who returned from service in the Gulf War with Gulf War syndrome and MCS, testified: "I, for one, would trade a shrapnel wound or a loss of a limb for these illnesses from which we suffer that are chronic, debilitating, and definitely life-altering." Such a statement must seem shocking to anyone who has not lived with MCS and does not realize the extent to which workplaces, stores, restaurants, public meetings, social occasions, places of worship, and schools are all inaccessible for the person with severe chemical sensitivity.

4. It is also important to investigate the overlap between MCS and chronic fatigue syndrome, fibromyalgia, and sick building syndrome (see Appendix 4). Substantial percentages of people with these other syndromes report that they have become very sensitive to chemical exposures. And many others undoubtedly have simply not yet made the connection since they are so immersed in chemical exposures that they cannot see individual cause-and-effect relationships because of the phenomenon called masking. (Dr. Claudia Miller describes masking in Appendix 3.)

The relationship of these other conditions to chemical exposures is not simply one of academic interest. The personal histories in this collection suggest that many people suffering from these other syndromes might well be able to eliminate at least some of their symptoms through avoiding exposures that are triggering those symptoms. Consider the case of Erica, for example, whose fibromyalgia disappeared when she eliminated dairy products from her diet. Michael's chronic fatigue was directly related to his exposure to pesticides. The pain that Louise had lived with for months vanished immediately when she stopped soaking in bath water contaminated with pesticide runoff.

5. There are two issues related to MCS that need immediate funding: research and housing. Anyone interested in contributing money for MCS research can contact any of the six physicians/researchers who appeared in my videos for further information about where research money could best be used.

Housing is the paramount issue for people with MCS—it's not easy for someone with chemical sensitivities to find a good place to live. There are far too many people who have built or bought what they hoped would be a house they could tolerate, often at great financial sacrifice, only to find that their neighbors routinely spray pesticides that make them very sick. Others find that they cannot tolerate some of the building materials, and in some cases residual fragrances left by prior owners of a house cannot be eliminated and cause severe symptoms. Many people with MCS have committed suicide because they were unable to find housing that did not make them terribly sick. As the personal histories in this collection illustrate all too well, many others across the country are living desperate lives. Some may be about to give up. Nancy and Abner might well be alive today had they been able to find suitable housing where they weren't exposed to unacceptable levels of toxic chemicals.

Finding safe housing is crucial for an individual attempting to climb out of the mire of MCS because living with constant exposure to toxic chemicals exacerbates or perpetuates the condition. Medications, including painkillers, are just another form of chemical exposure and often make MCS patients worse. Avoidance of chemical exposures is the one therapy that seems to help everyone with MCS feel better. In some cases a period of avoidance allows people with MCS an opportunity to reduce their level of sensitivity to chemicals sufficiently to enable them to work and move about more freely in society. With no other consistently efficacious treatment yet available for MCS, it is important to provide opportunity for avoidance by making suitable housing available. And if people can feel healthy at least in their own homes, they may be able to work out of their homes and get off disability.

One major housing problem is that few individuals can afford to buy a tract of land sufficiently large to serve as a buffer against neighboring farmers who use pesticides and herbicides on their crops or neighbors who use these toxic chemicals on their lawns or use fireplaces, wood stoves, or barbecues. Another difficult problem is that different people are sensitive to different chemicals, so building materials that are tolerated by one chemically sensitive person may make another sick. The ideal situation would be to purchase a large tract of land, build various model MCS homes from a variety of materials, and then make them available to rent. If someone found that a particular model worked well for a month, they could then with some confidence rent it long term or have a duplicate built elsewhere on the tract of land.

One added advantage of MCS rental housing projects is that such housing would afford an opportunity for people with fibromyalgia, chronic fatigue syndrome, asthma, or other chronic health problems to try living in a less-toxic environment for a few weeks or months to see if their condition improves.

Challenging as it is to construct good MCS housing, with sufficient sums of money, housing could be built that would at least be much better than what is currently available to those with MCS. I would like to hear from anyone who would be interested in contributing to or investing in "relatively safe" housing projects for people with MCS. In theory, such a project should be a good investment because so many people are desperately looking for such housing; offering it would be like selling life preservers to the drowning. If enough people come forward with offers of substantial funds, I will endeavor to see that appropriate foundations or business structures are set up to build and manage the housing. Please write me at the address below if you are willing to help:

Alison Johnson
MCS Information Exchange
2 Oakland Street
Brunswick, ME 04011

Appendix 1

Multiple Chemical Sensitivity Videos

The following videos may be ordered for $20 each, $3 S&H or $4 priority shipping. Volumes II and III can be ordered as a pair for $38 plus $4 S&H. Order from:

Alison Johnson
MCS Information Exchange
2 Oakland Street
Brunswick, ME 04011

Multiple Chemical Sensitivity
How Chemical Exposures
May Be Affecting Your Health
90 Minutes

This original video in the series illustrates the devastating effect that MCS has had on the lives of sixteen patients from various backgrounds. The people with stories in this volume who appear on the video are Alberta, Danielle, Erica, Jennifer, Karen, Louise, Michael, Moises, Randa, Richard, Tim, Timothy, Tomasita, Tony, and Zach. The following national experts also appear in this video:

Nicholas A. Ashford, Ph.D., J.D., is Professor of Technology and Policy at the Massachusetts Institute of Technology. He has previously served as Chair of the National Advisory Committee on Occupational Safety and Health and as Chair of the EPA Committee on Technology, Innovation, and Economics.

Iris R. Bell, M.D., Ph.D., is Associate Professor at the University of Arizona Health Sciences Center and is also a member of the staff at the Tucson Veterans Affairs Medical Center, Tucson, Arizona.

Gunnar Heuser, M.D., Ph.D., is a toxicologist in private practice and is also Assistant Clinical Professor of Medicine at the UCLA School of Medicine.

William J. Meggs, M.D., Ph.D., is Associate Professor in the Department of Emergency Medicine, East Carolina University School of Medicine.

Claudia S. Miller, M.D., M.S., is Associate Professor in Environmental and Occupational Medicine at the University of Texas Health Science Center at San Antonio. Dr. Miller is a member of the Department of Veterans' Affairs Gulf War Expert Scientific Advisory Committee.

Gerald Ross, M.D., until recently practiced environmental medicine at the Environmental Health Center, Dallas, Texas. He will be opening a clinic in Utah in 2000. Dr. Ross was the first director of a clinic in Halifax, Nova Scotia, that is the first government-sponsored clinic in the world established to evaluate and treat patients with chemical sensitivity.

Multiple Chemical Sensitivity
Volume II: Commentary by Three Experts

Claudia Miller, M.D., M.S.
Nicholas Ashford, Ph.D., J.D.
Iris Bell, M.D., Ph.D.
70 Minutes

Multiple Chemical Sensitivity
Volume III: Commentary by Four Experts

Gerald Ross, M.D.
William Meggs, M.D., Ph.D.
Gunnar Heuser, M.D., Ph.D.
Will Pape, Business Executive
82 Minutes

Volumes II and III are intended to be companion videos to the original video, *Multiple Chemical Sensitivity: How Chemical Exposures May Be Affecting Your Health*, which is the best introduction to the subject. A brief indication of the topics the various experts discuss on Volumes II and III follows.

Claudia Miller

- Toxicant-Induced Loss of Tolerance (TILT)
- Short-term memory loss and mood swings
- Service on VA Gulf War Scientific Expert Committee
- Chemical exposures during the Gulf War
- Chemical sensitivity among ill Gulf War vets
- Psychological symptoms resulting from physical causes
- Overlap of chronic fatigue syndrome, fibromyalgia, and MCS

Nicholas Ashford

- Need for event-driven research
- Environmental medical unit as a research tool
- Work as director of a nine-country European commission that studied MCS
- Wood preservative illness in Germany
- Development of animal models
- Need for regulation of organophosphate pesticides and certain organic solvents
- Need for accommodation of chemically sensitive workers
- Why traditional toxicologists and epidemiologists have trouble understanding chemical sensitivity
- Development of chemical sensitivity as a two-step process
- Effects of endocrine-disrupting chemicals
- Arguments against a psychogenic basis for chemical sensitivity
- Political reasons for lack of funding for research

Iris Bell

- Nervous system involvement in chemical sensitivity
- Discussion of limbic kindling
- Female animals in experiments more likely to sensitize than males
- Need for funding from federal government for research
- Discussion of pilot Gulf War syndrome study

Gerald Ross

- Dealing with skepticism in the medical community
- Gulf War syndrome
- Personal experience as a chemically sensitive patient
- Acceptance of chemical sensitivity by various governments in Canada
- Debate over whether chemical sensitivity is a psychological or physical condition
- SPECT and PET brain scans
- Sick building syndrome and worker efficiency

William Meggs

- Chemical irritant asthma and chemical irritant rhinitis
- Changes in nasal linings of MCS patients shown through biopsy —infiltration with lymphocytes, defects in tight junctions between cells, proliferation of nerve fibers
- Neurogenic inflammation
- Study of industrial workers exposed to a chlorine dioxide spill
- Female/male ratio high not only for MCS but also for diseases like lupus and rheumatoid arthritis
- Importance of environmental chemicals in diseases like lupus and rheumatoid arthritis
- Chemicals in the indoor environment
- Spanish toxic oil syndrome outbreak in the 1980s in Spain

Gunnar Heuser

- Medical problems of patients who have sustained chemical injury
- Loss of voice from chemical exposure
- SPECT and PET brain scans
- Low natural killer cell levels as indicator of chemical injury
- Study showing efficacy of vitamin C in increasing number of natural killer cells
- Proliferation of chemicals and resultant chemical mixtures
- Distinction between allergy and chemical sensitivity

Will Pape

- Description of a building constructed with less-toxic materials and the resulting 40% decrease in worker absenteeism

Appendix 2

Statistical Results of a Survey
of the Experience of
351 MCS Patients with 160 Therapies

Survey Conducted by Alison Johnson

Therapies surveyed include Neurontin, magnets, ozone, hydrogen peroxide, mercury amalgam removal, macrobiotic diet, juicing, enzyme potentiated desensitization (EPD), Diflucan, Ultra-Clear, homeopathy, Nambudripad desensitization, Total Body Modification, chelation, massage, transfer factor, biofeedback, melatonin, and Prozac.

In investigating various therapies, I attempted to maintain a healthy degree of skepticism. The booklets listed below include descriptions of some of these therapies, using quotations from their leading proponents and words of caution from experts with a differing viewpoint. Quotations are also included from people who had great success with a specific therapy or found it harmful. The articles suggest to readers what questions they should be asking about some of these therapies and where problems might lie.

This survey was an attempt to fill the need for a mechanism that would allow those with MCS to share information quickly about what treatments are helpful, what ones are potentially dangerous, and what ones may deplete their already strained finances with no benefit. The results are of necessity based upon anecdotal evidence. That's not ideal, but it's better than nothing when real scientific evidence is simply not yet available.

Lynn Lawson, author of *Staying Well in a Toxic World*, has referred to this survey as "that superbly done, indispensable, annotated survey of MCS treatments." The survey was cited three times in the 1998 edition of *Chemical Exposures: Low Levels and High Stakes* by Nicholas Ashford and Claudia Miller. The following is a sample line from the table of results:

	Total #	Effect unclear	Harmful	Didn't help	Slight help	Major help	Enormous help
Prozac	63	4 6.3%	35 55.6%	5 7.9%	9 14.3%	8 12.7%	2 3.2%

The following booklets may be purchased as a package at a reduced price: $14 for MCS patients experiencing severe financial difficulty, $18 for other MCS patients, and $25 for health care providers. Individual booklets or articles may be purchased separately at the prices listed. Please make checks payable to MCS Information Exchange.

SEPTEMBER 18, 1996, BOOKLET

Complete booklet (plus 9-page table of survey results)	39 pages	$10.00
Relocation	12 pages	$3.00
Enzyme Potentiated Desensitization (EPD)	6 pages	$1.50
Applied Kinesiology	4 pages	$1.00
Antidepressants and Other Psychotropic Drugs	3 pages	$1.00
NAET (Nambudripad Allergy Elimination Technique)	2 pages	$1.00

NOVEMBER 8, 1996, BOOKLET

Complete booklet	43 pages	$8.00
Emotional Reactions	11 pages	$3.00
Ozone Generators	5 pages	$1.50
Provocation/Neutralization	5 pages	$1.50
Porphyria	6 pages	$1.50
Relocation	3 pages	$1.00
Mycoplasma Fermentans	5 pages	$1.50

MARCH 20, 1997, BOOKLET

Complete booklet	27 pages	$8.00
Neurontin	10 pages	$3.00
Mycoplasma Fermentans	9 pages	$3.00

SEPTEMBER 19, 1997, BOOKLET

Complete booklet (primarily about Neurontin)	13 pages	$4.00

FEBRUARY 13, 1998, BOOKLET

Complete booklet	19 pages	$6.00
Sauna-Detox	6 pages	$1.50
Neurontin	2 pages	$1.00
Antibiotics	2 pages	$1.00
Anesthesia	4 pages	$1.50

Appendix 3

Gulf War Syndrome

Committee on Veterans' Affairs
Subcommittee on Benefits
United States House of Representatives
October 26, 1999
Invited Testimony by Claudia S. Miller, M.D., M.S.

I have been asked to explain how physicians who see sick Gulf War veterans can observe the same or similar symptoms and interpret them as either undiagnosed illness or diagnosed illness. Even when doctors apply monikers to these patients' illnesses, like depression, migraine headaches, asthma, irritable bowel or fibromyalgia, these monikers do not explain why these veterans are sick. Most have symptoms involving several organ systems simultaneously. For them there is no unifying diagnosis offered, no etiology specified, and no disease process clarified.

In truth, all of these veterans are undiagnosed because what we are dealing with is an entirely new mechanism of disease not covered by standard medical diagnoses—one which presents itself symptomatically as different conditions to different specialists.

- The rheumatologist observing diffuse muscle pain diagnoses myalgias.

- The neurologist hearing head pain and nausea diagnoses migraine headaches.

- The pulmonologist finding airway reactivity diagnoses asthma.

- The psychiatrist seeing chronic malaise diagnoses depression.

- The gastroenterologist noting GI complaints diagnoses irritable bowel syndrome.

- Some private practitioners diagnose multiple chemical sensitivity, or MCS, which is not a diagnosis in itself, but rather just another manifestation of the underlying disease process.

So what is at the core of this myriad of symptoms that has come to be called "Gulf War syndrome"? What is the underlying disease process? The key is in the new-onset intolerances these people share.

Over the past six years, I have served as a consultant to the VA's referral center for Gulf War veterans in Houston. The vast majority of the veterans there reported multiple new intolerances since the War. Among the first 59 patients, 78% reported new-onset chemical intolerances; 40% experienced adverse reactions to medications; 78% described new food intolerances; 66% reported that even a can of beer made them feel ill; 25% became ill after drinking caffeinated beverages; and 74% of smokers felt sick if they smoked an extra cigarette or borrowed someone else's stronger brand. More than half reported new intolerances in all three categories—chemical inhalants, foods, and drugs or food/drug combinations. One mechanic said that before the Gulf War his idea of the perfect perfume was WD-40. Since the war, WD-40 and a host of other chemicals make him feel ill. Many veterans no longer fill their own gas tanks because the gasoline vapors make them "spacy" or sick. Some won't drive because they become disoriented in traffic and they fear causing an accident. Or they can't find their cars, forget where they are going or get lost in once familiar areas. One VA study found excess motor vehicle deaths among Gulf veterans and interpreted this as possible increased risk-taking behavior (Kang and Bullmann, 1996). What the veterans tell me is that they get confused, go off the road, mistake the accelerator for the brake, and have trouble judging stopping distances when they are exposed to gasoline, diesel exhaust, or freshly tarred roads. Researchers at the Robert Wood Johnson Medical School in New Jersey and at the University of Arizona have noted similar multi-system symptoms and intolerances to common chemicals, foods, and drugs among the veterans (Fiedler et al, 1996; Bell et al, 1998). And a CDC study found that ill Gulf War veterans reported more chemical intolerances than healthy veterans (Fukuda et al, 1998).

These studies are confounded by a phenomenon called "masking," which occurs when people become intolerant to many different things (Miller and Prihoda, 1999a). As they go through a day, symptoms triggered by fragrances, hair spray, vehicle exhaust, foods and medications pile up so they feel sick most of the time. No one cause can be isolated because there's too much background noise, and patients often underestimate the number of exposures that affect them.

This problem is not altogether new. German researchers described similar intolerances in chemical weapons workers after World War II (Spiegelberg, 1961). Nearly 20 percent of agricultural workers on a California registry for organophosphate pesticide poisoning (Tabershaw and Cooper, 1966) reported that even a "whiff" of pesticide made them sick with symptoms like those of the Gulf War veterans, as did dozens of government workers a decade ago, after the EPA headquarters became a "sick building" following remodeling (EPA, 1989). Similar outbreaks of chemical intolerances have been reported in more than a dozen countries (Ashford et al, 1995).

These observations suggest that we may indeed be dealing with an entirely new mechanism for disease, one which has been referred to with the acronym "TILT," or "Toxicant-Induced Loss of Tolerance" (Miller, 1996, 1997, 1999). Any one toxicant appears capable of initiating this process. TILT involves two steps, initiation and triggering (Ashford and Miller, 1998): (1) First, a single acute or multiple low-level exposures to a pesticide, solvent or other chemical causes loss of tolerance in a subset of those exposed; (2) Thereafter very low levels of common substances can trigger symptoms—not only chemicals, but various foods, medications, alcoholic beverages and caffeine. Symptoms involve several organ systems. These intolerances are the hallmark of TILT, just as fever is the hallmark symptom of infectious diseases.

Over the past several years, the finger has been pointed at a number of potential causes for Gulf War syndrome—everything from the oil shroud to pesticides, vaccinations, and pyridostigmine bromide. What set off the Gulf War veterans? The answer is "all of the above." Exposure to any one or any combination of these toxicants may, in fact, be capable of causing a general breakdown in tolerance that can result in a plethora of beguiling symptoms.

We do not know exactly how this breakdown in tolerance occurs. We do know that rats with nervous systems sensitive to organophosphate pesticides are also intolerant of diverse drugs and have increased gut permeability which in humans is associated with food intolerance (Overstreet et al, 1996). This suggests the breakdown might involve the cholinergic nervous system, which regulates processes throughout the body.

How can these people be helped? No one knows—yet. The biggest obstacle is the symptoms themselves, which serve as red herrings, diverting attention away from the central problem. What we do know is that Gulf War veterans who have come to recognize what sets them off and then avoid these triggers tend to improve. We need to apply this understanding to the diagnosis and treatment of other such veterans.

The first thing that needs to be done is to set up unmasking studies in which sick Gulf War veterans can be isolated from the exposures that are setting them off. This can be achieved by putting them in a special environmentally controlled hospital unit (Miller, 1997; Miller et al., 1997). Once we get them to baseline, we can reintroduce things like caffeine, perfumes, various foods, etc., and identify some of the things that cause their flare-ups. With avoidance, it is hoped that they, too, can improve. This combined diagnostic-therapeutic approach would eliminate much of the confusion that is the focus of this hearing. There is no simple answer to Gulf War illness. No single toxicant is likely to have caused it. But if we concentrate less on the original toxicants and more on the underlying disease mechanism, I believe we can make progress in understanding why these people are sick and what we can do to help them.

BIOSKETCH: Claudia S. Miller, M.D., M.S., is an Associate Professor in Environmental and Occupational Medicine in the Department of Family Practice of the University of Texas Health Science Center at San Antonio. Dr. Miller is co-author of the WHO-award-winning New Jersey Report on Chemical Sensitivity and a professionally acclaimed book, *Chemical Exposures: Low Levels and High Stakes*. Dr. Miller has served as a consultant to the Houston VA Regional Referral Center for Gulf War Veterans since 1993. She is also principal investigator on a study of neurobehavioral sensitization funded by the Office of Naval Research.

Editor's note: For a copy of Dr. Miller's nine-page article describing in detail her TILT theory and the phenomenon of masking, which appeared in *Environmental Health Perspectives* (see reference below), send $2 to MCS Information Exchange, 2 Oakland Street, Brunswick, ME 04011. In her discussion of masking, Dr. Miller explains why exposure chambers are flawed as research tools for establishing the existence or nonexistence of chemical sensitivity.

Solvents
- Glues
- Paints
- Gasoline
- Nail polish/remover

Indoor air volatile organic compounds
- New carpet
- Plasticizers
- Formaldehyde
- Fragrances

Drugs/medical devices
- Vaccines
- Anesthetics
- Implants
- Antibiotics

Toxicant-induced Loss of Tolerance

Combustion-related products
- Engine exhaust
- Tobacco smoke
- Oil well fire smoke
- Natural gas
- Tar/asphalt

Pesticides
- Organophosphates
- Carbamates, pyridostigmine
- Pentachlorophenol
- Pyrethrins

Cleaners
- Phenolic disinfectants
- Ammonia
- Bleach

Figure 1. Exposures that may initiate TILT or trigger symptoms

Ear, nose and throat
- Sinusitis
- Polyps
- Tinnitus
- Recurrent otitis

Neuropsychological
- Multiple Chemical Sensitivity
- Attention Deficit Hyperactivity Disorder (ADHD)
- Depression/manic-depression
- Migraines and other headaches
- Seizures

Cardiovascular
- Arrhythmias
- Hypertension
- Hypotension
- Raynaud's phenomenon

Miscellaneous
- Chronic Fatigue Syndrome
- Implant-associated illness
- "Gulf War Syndrome"

Toxicant-induced Loss of Tolerance

Respiratory
- Asthma
- Reactive Airways Dysfunction Syndrome (RADS)
- Toluene Diisocyanate (TDI) Hypersensitivity

Skin
- Eczema
- Hives
- Other rashes, eruptions

Connective tissue/musculoskeletal
- Fibromyalgia
- Carpal tunnel syndrome
- Temporomandibular Joint Dysfunction (TMJ) Syndrome
- Arthritis
- Lupus

Gastrointestinal
- Irritable bowel
- Reflux

Figure 2. Conditions that may have their origins in TILT

References

AGENCY FOR TOXIC SUBSTANCES AND DISEASE REGISTRY (ATSDR) (1994). Proceedings of the Conference on Low Level Exposure to Chemicals and Neurobiologic Sensitivity. *Tox. Ind. Health* 10(4/5):25.

ASHFORD, N., HEINZOW, B., LTJEN, K., MAROULI, C., MLHAVE, L., MNCH, B., PAPADOPOULOS, S., REST, K., ROSDAHL, D., SISKOS, P., and VELONAKIS, E. (1995). "Chemical Sensitivity in Selected European Countries: An Exploratory Study." Ergonomia Ltd., Athens, Greece.

ASHFORD, N., and MILLER, C. (1998). *Chemical Exposures: Low Levels and High Stakes.* New York, Wiley & Sons.

BELL, I., WALSH, M., GROSS, A., GERSMEYER, J., SCHWARTZ, G., and KANOF, P. (1997). "Cognitive Dysfunction and Disability in Geriatric Veterans with Self-Reported Intolerance to Environmental Chemicals." *J. Chronic Fatigue Syndrome.* 3(3):15- 42.

ENVIRONMENTAL PROTECTION AGENCY (EPA) (1989). Report to Congress on Indoor Air Quality, Volume II, Assessment and Control of Indoor Air Pollution.

FIEDLER, N., KIPEN, H., NATELSON, B., and OTTENWELLER, J. (1996). "Chemical Sensitivities and the Gulf War: Department of Veterans Affairs Research Center in Basic and Clinical Science Studies of Environmental Hazards." *Regulatory Tox. Pharmacol.* 24: S129-S138.

FUKUDA, K., NISENBAUM, R., STEWART, G., THOMPSON, W., ROBIN, L., WASHKO, R., NOAH, D., BARRETT, D., RANDALL, B., HERWALDT, B., MAWLE, A., and REEVES, W. (1998). "Chronic Multi-system Illness Affecting Air Force Veterans of the Gulf War." *JAMA* 280: 981-988.

KANG, H., and BULLMAN, T. (1996). "Mortality Among U.S. Veterans of the Persian Gulf War." *New Engl. J. Med.* 335(2a):1498-1504.

MILLER, C. (1996). "Chemical Sensitivity: Symptom, Syndrome or Mechanism for Disease?" *Tox.* 11: 69-86.

MILLER, C. (1997). "Toxicant-Induced Loss of Tolerance-An Emerging Theory of Disease?" *Environ. Health Perspect.* 105 (Suppl. 2): 445-453.

MILLER, C. (1999). "Are We on the Threshold of a New Theory of Disease? Toxicant-induced Loss of Tolerance and its Relationship to Addiction and Abdiction" *Tox. Ind. Health.* 15:284-294.

MILLER, C., and PRIHODA, T. (1999a). "A Controlled Comparison of Symptoms and Chemical Intolerances Reported by Gulf War Veterans, Implant Recipients and Persons with Multiple Chemical Sensitivity." *Tox. Ind. Health* 15:386-397. MILLER, C., and PRIHODA, T. (1999b). "The Environmental Exposure and Sensitivity Inventory (EESI): A Standardized approach for measuring Chemical Intolerances for Research and Clinical Applications." *Tox. Ind. Health* 15:370-385.

MILLER, C., ASHFORD, N., DOTY, R., LAMIELLE, M., OTTO, D., RAHILL, A., and WALLACE, L. (1997). "Empirical Approaches for the Investigation of Toxicant-Induced Loss of Tolerance." *Environ. Health Perspect.* 105 (Suppl. 2): 515-519.

OVERSTREET, D., MILLER, C., JANOWSKY, D., and RUSSELL, R. (1996). "Potential Animal Model of Multiple Chemical Sensitivity with Cholinergic Supersensitivity." *Tox.* 111: 119-134.

SPIEGELBERG, V. (1961). "Psychopathologisch-neurologische Sch_den nach Einwirkung Synthetischer Gifte." In *Wehrdienst und Gesundeir,* Vol. III. Darmstadt: Wehr und Wissen Verlagsgessellshaft (1961).

TABERSHAW, I., and COOPER, C. (1966). "Sequelae of Acute Organic Phosphate Poisoning" *J. Occup. Med.* 8:5-20.

Gulf War Veterans' Illnesses:
Second Report by the
House Committee on Government
Reform and Oversight
105th Cong., 1st Sess., House Report 105-388, November 7, 1997

Excerpts from this report are printed below (the square brackets and the material they enclose appeared in the House report). The report describes research on the nerve gas sarin that the Army asked Dr. Frank Duffy, associate professor of neurology at Harvard University Medical School, to conduct in the 1970s.

> The results, according to Dr. Duffy, indicated that "low levels of exposure to the nerve agent sarin can produce long-lasting effects. It was perfectly clear that not only were people, after [low level sarin] exposure showing long-term effects, but it was widely accepted in the pesticide industry that exposure to related compounds like malathion and parathion or the chlorinated hydrocarbon insecticides led to long-term consequence." . . .
>
> In a 1987 letter to Robert Hall of the Hawaii Institute for Biosocial Research, Dr. Duffy also noted the possible confusion between organophosphate-delayed-neuropathy and stress: "I applaud your effort in raising the level of consciousness about the serious potential for long-term effects due to exposures to these [organophosphate] compounds. It has been our experience that the side effects of minimal but continual exposures to the compounds *mimic the symptoms associated with a stressful life* [emphasis added]. Accordingly most individuals are unable to determine whether their irritability is related to a stressful life or to a recent organophosphate exposure. This is a serious issue."

The House committee report also mentions the work of Robert Haley, M.D.:

> Robert Haley and his research colleagues at the University of Texas Southwestern Medical Center also completed a study in early 1997 on Gulf veterans, the results of which were published in three articles in JAMA. According to the study, "Some Gulf War veterans may have delayed, chronic neurotoxic syndromes from wartime exposure to combinations of chemicals" and "clusters of symptoms of many Gulf War veterans reflect a

spectrum of neurologic injury involving the central, peripheral, and autonomic nervous systems."[1]

With respect to the issue of why not all Gulf War vets came home sick when so many did, the House report states:

> People have asked why most Gulf War veterans have not reported illnesses while only some veterans were affected. Dr. Kenneth Olden, director of the National Institute of Environmental Health Sciences, was recently quoted in the press: "We've known for a long time that when several hundred people are exposed to the same environmental toxicants, some people get sick and others don't. There are a number of enzyme systems that detoxify chemicals. If you have too little—that's a problem."

A December 1, 1999, article in the *San Francisco Chronicle* states: "New research into Gulf War illness released yesterday has found the first evidence of brain damage in ailing veterans, lending further support to suspicions that toxic chemicals are the cause of a mysterious affliction that has affected more than 30,000 former U.S. troops."

Researchers at the University of Texas Southwestern Medical Center in Dallas used magnetic resonance spectroscopy to show that a group of 22 vets with Gulf War syndrome had concentrations of a brain chemical called N-acetyl-aspartate that range from 10 to 25 percent lower than in a group of healthy veterans. Dr. James Fleckenstein, one of the researchers, said that even a 10 percent loss of this chemical would amount to a "pretty severe hit" to brain function. The *Chronicle* also quotes Fleckenstein as saying, "This validates that these are sick people, not people who are crazy, or depressed or trying to get money." According to the article, the researchers suggest that the lower amount of this key chemical indicates a "loss of neurons in the brain stem, which controls some body reflexes, and in the basal ganglia, which is the brain's switching station for movement, memory and emotion."

---

[1] In the June 16, 1999, issue of *Toxicology and Applied Pharmacology*, Dr. Haley and other researchers published results of a study showing that Gulf War vets who developed Gulf War syndrome have lower levels of an enzyme known as PON-Q than Gulf War vets who did not become sick. PON-Q acts to detoxify several organophosphates, including the nerve agent sarin. (Many leading pesticides are organophosphates and act as neurotoxins.)

Appendix 4

Overlap Between
Multiple Chemical Sensitivity (MCS)
Chronic Fatigue Syndrome (CFS)
Fibromyalgia (FMS)
Gulf War Syndrome (GWS)

1994 CDC Criteria for CFS:

1. Clinically evaluated, unexplained, persistent, or relapsing chronic fatigue that is of new or definite onset (has not been lifelong); is not the result of ongoing exertion; is not substantially alleviated by rest; and results in substantial reduction in previous levels of occupational, educational, social, or personal activities.

2. Concurrent occurrence of four or more of the following symptoms, all of which must have persisted or recurred during six or more consecutive months of illness and must not have predated the fatigue:

   a. self-reported impairment in short-term memory or concentration severe enough to cause substantial reduction in previous levels of occupational, educational, social or personal activities
   b. sore throat
   c. tender cervical or axillary lymph nodes
   d. muscle pain
   e. multijoint pain without swelling or redness
   f. headaches of a new type, pattern or severity
   g. unrefreshing sleep
   h. postexertional malaise lasting more than 24 hours.

From *Annals of Internal Medicine* 121, no. 12 (December 14, 1994), Table 14-1.

American College of Rheumatology's Criteria for the Diagnosis of Fibromyalgia (1990)

1. A 3-month or longer history of widespread pain (above and below the waist, on the right and left sides, and axially) and

2. pain on direct palpation of 11 of 18 tender point sites.

From Wolfe F., Smythe, H.A., Yanus, M.B., et al., "Report of the Multicenter Criteria Committee," *Arthritis Rheum* 33 (1990):160-72.

———————————

Grace Ziem, M.D. (Baltimore, Maryland), who specializes in treating chemical injury, reports that 85% of her MCS patients meet the CDC's criteria for chronic fatigue syndrome and 67% have fibromyalgia. Albert Donnay, executive director of MCS Referral & Resources, discusses the issue in a 1997 paper titled "Overlapping Disorders: Chronic Fatigue Syndrome, Fibromyalgia Syndrome, Multiple Chemical Sensitivity, & Gulf War Syndrome." According to Donnay, "Which of these diagnoses a person receives usually depends on the type of specialist he or she sees. CFS [chronic fatigue syndrome] is most likely to be diagnosed by infectious disease specialists, FMS [fibromyalgia] by rheumatologists, MCS [multiple chemical sensitivity] by occupational and environmental medicine physicians, and GWS [Gulf War syndrome] by physicians in Veterans' Affairs hospitals."

Appendix 5

Research Studies and Articles

In the following small sampling of articles on toxic chemicals, *EHP* stands for *Environmental Health Perspectives*, published by the National Institute of Environmental Health Sciences, a division of the NIH.

"Exposures from Indoor Spraying of Chlorpyrifos Pose Greater Health Risks to Children than Currently Estimated," D.L. Davis and A.K. Ahmed, *EHP* 106, no. 6 (June 1998): 299-304:

> Recent studies by Gurunathan and her associates at the Environmental and Occupational Health Sciences Institute at Rutgers University (the EOHSI study) indicate that broadcast spraying of chlorpyrifos in the indoor environment may pose considerable risk to public health. In this well-conducted study, applications of chlorpyrifos by trained applicators following recommended procedures produced pesticide residues on children's toys and on hard surfaces in test rooms approximately 21-119 times above the recommended reference dose (RfD) of 3 µg/kg/day for chlorpyrifos exposure to children from all sources. . . .
>
> The EOHSI study demonstrated that the compound continued to be released into the gas phase and became deposited on a variety of solid surfaces for at least 2 weeks after a single broadcast application. In accord with the manufacturers' recommended practices for a standard apartment style room, no toys were in the test rooms during the spraying period. Children's toys (consisting of plush and plastic materials) were placed in the test rooms 1 hr after spraying, and measurements of the accumulation of chlorpyrifos residues on these toys were made on days 1 through 14. The EOHSI study showed that these commonly used children's toys contained consistently high concentration[s] of chlorpyrifos residues over a 2-week period, thus serving both as a chemical sink and a long-term reservoir for the toxic compound.

"Exposures of Children to Organophosphate Pesticides and Their Potential Adverse Health Effects," B. Eskenazi, A. Bradman, and R. Castorina, *EHP* 107, supplement 3 (June 1999): 409-16:

Tests on young rodents demonstrate a progressive decrease in susceptibility to OP [organophosphate] pesticides with increasing age. In some cases, the lethal dose in immature animals is only 1% of the adult lethal dose. . . . In humans, children have had higher fatality rates than adults in several cases of OP poisoning. . . .

According to the National Academy of Sciences, children's OP exposures are of special concern because "exposure to neurotoxic compounds at levels believed to be safe for adults could result in permanent loss of brain function if it occurred during the prenatal and early childhood period of brain development."

"Pregnancy Outcome Following Gestational Exposure to Organic Solvents," S. Khattak, et al., *JAMA* 281, no. 12 (March 24/31, 1999) : 1106-1109:

Occupational exposure to organic solvents during pregnancy is associated with an increased risk of major fetal malformations. This risk appears to be increased among women who report symptoms associated with organic solvent exposure. Women's exposure to organic solvents should be minimized during pregnancy.

"The Stuff in the Backyard Shed," *U.S. News and World Report*, November 8, 1999. This article on chlorpyrifos, "marketed . . . under the names Dursban (for structures) and Lorsban (for agriculture)," states that it:

poses a particular threat to the developing nervous system, attacking it in ways that can lower intelligence and cause behavioral problems. . . .

New research supports the notion that there is a hypersensitive minority, susceptible to quantities of chlorpyrifos that wouldn't faze the average person. . . . Is this substance short-circuiting some people's brains even as it snuffs out bugs?

"Environmental Health," *EHP* 106, no. 12 (December 1998): A600-A603:

One of the most pressing environmental health priorities for the developed world in the 21st century is posed by the thousands of industrial chemicals for which even the most basic test data are not available. According to the EPA report *Chemical Hazard Data Availability Study,* released in April 1998, only 7% of the 3,000 high production volume chemicals (HPVs) currently used in commerce in the United States have a comprehensive set of basic health and environmental effects studies available.

Appendix 6

Resources

Ashford, Nicholas A., Ph.D., J.D., and Claudia S. Miller, M.D., M.S. *Chemical Exposures: Low Levels and High Stakes*. 2nd ed., 428 pages. John Wiley, 1998. $39.95. 1-800-225-5945. The *Journal of the American Medical Association* has written: "Clinicians and policymakers would do well to read and heed the advice of this book."

Lawson, Lynn. *Staying Well in a Toxic World: Understanding Environmental Illness, Multiple Chemical Sensitivities, and Sick Building Syndrome.* 480 pages. Lynnword Press. To order, send a check for $18.95 (includes S&H) to Lynn Lawson at Staying Well, P.O. Box 1732, Evanston, IL 60201.

McCampbell, Ann, M.D. *What Is MCS?* 16 pages. To order, send a check for $4 to Ann McCampbell, M.D., 13 Herrada Road, Santa Fe, NM 87505.

Rapp, Doris, M.D. *Is This Your Child?* Also very useful for adults. William Morrow, 1991. 655 pages, $12. Available in most bookstores.

Lists of physicians interested in treating patients with MCS can be obtained from the following organizations:

American Academy
of Environmental Medicine
7701 East Kellogg, Ste. 625
Wichita, KS 67207
316-684-5500

Association of Occupational
and Environmental Clinics
1010 Vermont Ave., NW, Ste. 513
Washington, D.C. 20005
202-347-4976

Addresses for various MCS support groups and newsletters can be obtained from the following national MCS organizations (or send $1 and a SASE to MCS Information Exchange, 2 Oakland St., Brunswick, ME 04011, for a more complete list):

Human Ecology Action League
P.O. Box 29629
Atlanta, GA 30359-0629
404-248-1898

Chemical Injury Information Network
P.O. Box 301
White Sulphur Springs, MT 59645-0301
406-547-2255

Appendix 7

Useful Groupings of Contributors

Individuals who consider the listed substance a major contributing factor in developing MCS:

Pesticide: Michael, Erica, Louise, Diane, Marie, Ariel, Tomasita, Nancy, Janice, Jacob, Danielle, Bob, Effie, Nicole, Joy, Ambrose, Eli

Paint: Tony, Richard, Jane, Al, Jennifer

New Carpet: Jim, Randa, Timothy, Bonnie, Carina, John, Carole, Karen

Sick Building: Erica, Marilyn, Lizbeth, Diane, Tina, Randa, Liz, John, Sandy, Carole, Effie, Karen

Renovated Building: Randa, Bonnie, Carina, John, Karen

Individuals who discuss the stated topic in their personal histories:

Housing problems: Michael, Erica, Tony, Zach, Louise, Jim, Diane, Marie, Ariel, Donna, Tomasita, Randa, Abner, Bonnie, Carina, Nancy, Janice, John, Jacob, Carole, David, Ann, George, Rand, Nicole, Joy, Jennifer, Karen, Eli

Fragrance exposure problems: Michael, Terry, Marilyn, Tim, Linda, Tony, Carl, Diane, Marie, Richard, Donna, Tomasita, Randa, Abner, Janice, Sandy, Bob, Effie, David, Ann, Rand, Jennifer, Alberta, Karen, Alison

Brain fog and memory problems: Michael, Erica, Pat, Kelly, Carl, Diane, James, Marie, Connie, Richard, Randa, Rand, Al, Karen

Anger arising from chemical exposures: Michael, Jeff, Kelly, Zach, Jim, Tomasita, Moises, Jane, Rand

Appendix 8

Ronald Gots and the
Environmental Sensitivities Research Institute

In their book titled *Chemical Exposures: Low Levels and High Stakes,* Nicholas Ashford, Ph.D., J.D., and Claudia Miller, M.D., M.S., describe the Environmental Sensitivities Research Institute as follows:

> ESRI is a corporate-supported entity with an "Enterprise Membership" fee of $10,000 per year. Board members include DowElanco; Monsanto; Procter and Gamble; the Cosmetic, Toiletry and Fragrance Association; and other companies and trade associations involved in the manufacture of pharmaceuticals, pesticides, and other chemicals.[1]

Ashford and Miller quote this passage from an article that ESRI paid a public relations firm to circulate to newspapers across the country:

> Scientists are increasingly concerned that a doubtful new diagnosis—supposedly caused by everything "man made" in the environment—is unnecessarily making thousands of Americans miserable each year. One of these so called "modern diseases" is called MCS, for Multiple Chemical Sensitivities. Many established scientists and physicians doubt MCS actually exists; it exists only because a patient believes it does and because a doctor validates that belief.[2]

Until recently, when he incurred the wrath of a district court judge and was criticized in front-page articles in the *Wall Street Journal* and the *Washington Post,* Ronald Gots was the director of ESRI. As Peter

---

[1] Nicholas A. Ashford and Claudia S. Miller, *Chemical Exposures: Low Levels and High Stakes* (New York: John Wiley, 1998), p. 279, n1.

[2] Ibid.

Radetsky notes in *Allergic to the Twentieth Century*, "As one of the most vocal of those who insist on the psychogenic origin of MCS, Gots provides the chemical industry just what it needs: public assurance that whatever else might be going on in MCS, chemicals have nothing to do with it."[3] When John Stossel presented an extremely negative view of MCS on *20/20*, who was the expert to whom he turned for a medical opinion about MCS? Ronald Gots of course.

Gots also ran until recently a company called Medical Claims Review Services, which like ESRI, was based in Bethesda, Maryland. In a front-page *Washington Post* article (July 4, 1999) titled "Cutting Claims with Fraud," reporter Edward Walsh described how a leading automobile insurer was cutting costs through a process called "utilization review." The insurance company in question had used Ronald Gots's Medical Claims Review Services as part of the process of determining whether insurance benefits should be awarded. The *Post* quotes a 1998 opinion by Idaho District Judge D. Duff McKee in a lawsuit related to the insurer's conduct. Judge McKee stated that "the evidence was overwhelming that the utilization review company selected by the claim examiner was a completely bogus operation. The company did not objectively review medical records but rather prepared 'cookie-cutter' reports of stock phrases, assembled on a computer, supporting the denial of claims by insurance companies. The insured's medical records were not examined and reports were not prepared by doctors or even reviewed by doctors."

During the years that he was running ESRI, Ronald Gots was an extremely active point man, traveling to various sites in the country where it appeared MCS was gaining credence. When MCS advocates in New Mexico were having some success in interesting members of the state legislature in introducing MCS bills, Gots spent several days in the state. A staff member of the New Vistas Independent Living Center in Santa Fe, an organization that provides support for people with disabilities, told me that Gots requested a meeting with staff members and argued that they should not be running a support group for people with MCS. Gots's other activities in New Mexico included a meeting with members of a Medicaid advisory committee at which he argued his position that MCS is a psychologically based illness.

Gots and Cindy Lynn Richard, who is currently on the staff of ESRI, were also involved in the production of a booklet titled *A Close Look at*

---

[3] Peter Radetsky, *Allergic to the Twentieth Century* (Boston: Little, Brown & Company, 1997), p. 180.

Multiple Chemical Sensitivity, which was written by Stephen Barrett, M.D., a psychiatrist, and published by Quackwatch, Inc. On the inside of the back cover is a statement thanking Gots and Richard for "valuable help with this report." In Appendix C, "Reputable Consultants," Gots is one of five persons listed as scientific experts, and Richard is the sole entry for a public policy consultant. Dr. Barrett and Dr. Gots gave a joint presentation of the position that MCS is a psychologically based condition in a debate on chemical sensitivity sponsored by the American College of Toxicologists in Florida in November 1998.

On page 40 of the booklet, Dr. Barrett urges state licensing boards to "scrutinize the activities of clinical ecologists" (physicians practicing environmental medicine used to refer to themselves as clinical ecologists) and states his belief that "most of them should be delicensed." The part of the booklet that will particularly dismay anyone struggling with chemical sensitivity is the four pages titled "Recommendations" at the end of the booklet. Implementing certain of these suggestions would be tantamount to stepping on the fingers of someone about to lose their grasp on the ledge of a skycraper. Among Dr. Barrett's recommendations are the following:

- Psychiatrists and psychologists should not "reinforce unsubstantiated beliefs about MCS" and should assist those with MCS to "restructure their beliefs."

- Physicians should persuade patients with MCS to seek help from those in the mental health field.

- The National Institute of Environmental Health Sciences, a subdivision of the NIH, should cease putting out information that "suggests that MCS is a clearly defined disease entity caused by exposure to environmental chemicals."

- Legislators should not provide money for MCS research that has "no practical value."

- Relatives of patients who come "under the spell of a clinical ecologist" should take quick measures to protect themselves from "financial ruin."

• Educators should not provide home tutoring or other accommodations for children with MCS because such actions will give children "false messages . . . about their health status."

These words do not exist in a vacuum—they can have a devastating effect on people's lives. Consider the impact the above recommendations to relatives could have on the lives of all those people with MCS who are living in poverty, barely able to survive on meager disability payments, if they were even lucky enough to obtain those.

Think of Zach when you consider the recommendation to educators. The school system that has been providing him with home tutors and letting him participate in outdoor recess and physical education is currently attempting to withdraw those services. Zach's mother is very concerned that Zach may lose not only the chance to interact with his adult tutors but also the chance to play with other children.